Biodiversity in India

Volume 3

T. Pullaiah

2024
Regency Publications
A Division of
Astral International Pvt. Ltd.
New Delhi – 110 002

First Published, 2004
Reprinted, 2024

ISBN: 9788187498940 (HB)

Publisher's Note:

Every possible effort has been made to ensure that the information contained in this book is accurate at the time of going to press, and the publisher and author cannot accept responsibility for any errors or omissions, however caused. No responsibility for loss or damage occasioned to any person acting, or refraining from action, as a result of the material in this publication can be accepted by the editor, the publisher or the author. The Publisher is not associated with any product or vendor mentioned in the book. The contents of this work are intended to further general scientific research, understanding and discussion only. Readers should consult with a specialist where appropriate.

Every effort has been made to trace the owners of copyright material used in this book, if any. The author and the publisher will be grateful for any omission brought to their notice for acknowledgement in the future editions of the book.

Published by : **Regency Publications**
A Division of
Astral International Pvt. Ltd.
– ISO 9001:2015 Certified Company
4736/23, Ansari Road, Darya Ganj
New Delhi-110 002
Ph. 011-43549197, 23278134
E-mail: info@astralint.com
Website: www.astralint.com

PREFACE

Since Rio Earth Summit, Biodiversity has gained importance. Biodiversity bill has been approved by the Parliament of India. Several countries are preparing national Biodiversity Strategy and Action Plans. Even after this not many journals and books are coming out on Biodiversity in India. Flora of India is taking time for completion, only Volumes 1-5, 12 and 13 have come out. We intend to bring out a series of Edited volumes on Biodiversity in India encompassing all aspects. This includes species diversity, Genetic diversity and Ecosystem diversity. This volume is the third in this attempt. Authors who intend to publish their articles are requested to send them to the editor. We seek the opinion of the readers for the coming volumes.

CONTENTS

LIST OF CONTRIBUTORS

Prof. T. Pullaiah
Dean, Faculty of Life Sciences
Department of Botany
Sri Krishnadevaraya University
Ananatapur – 515 003
ANDHRA PRADESH

Prof. G.S. Rawla
Department of Botany
Punjab University
Chandigarh – 160 014

Dr. S.P. Gaikwad
Department of Botany
Shivaji University
Kolhapur – 416 004
MAHARASHTRA

Prof. S.R. Yadav
Department of Botany
Delhi University
Delhi – 110 007

Prof. K.V. Mallaiah
Department of Botany and
Microbiology
Nagarjuna University
Nagarjuna Nagar – 522 510
ANDHRA PRADESH

Dr. M. Vijaya Koteswari
Lecturer
Hindu College
Guntur – 522 001

K.P.S. Brar
Department of Botany
Punjab University
Chandigarh – 160 014

Prof. T.N. Mary
Department of Botany
Nagarjuna University
Nagarjunna Nagar – 522 510
Guntur (Dt.)
ANDHRA PRADESH

Prof. J.B. Atluri
Department of Botany
Andhra University
Visakhapatnam – 530 003
ANDHRA PRADESH

Dr. S.P. Venkata Ramana
Department of Botany
Andhra University
Visakhapatnam – 530 003
ANDHRA PRADESH

Prof. C. Subba Reddi
Department of Environmental Sciences
Andhra University
Visakhapatnam – 530 003
ANDHRA PRADESH

Prof. B. Hunumanta Rao
Department of Botany
Andhra University
Visakhapatnam – 530 003
ANDHRA PRADESH

M.N.V. Peddi Naidu
Department of Botany
Andhra University
Visakhapatnam – 530 003
ANDHRA PRADESH

CHAPTER 1

CLITOCYBE KUMMER & *LEPISTA* (FR.) W.G. SMITH (TRICHOLOMATACEAE: AGARICALES) IN INDIA

*G.S. Rawla and K.P.S. Brar**

Thirteen species of *Clitocybe* and eight of *Lepista* have been so far reported from India by various workers (Sathe and Daniel, 1980; Sathe and Deshpande, 1980; Sathe and Kulkarni, 1980; Manjula, 1983; Rawla and Arya, 1987; Verma *et al.*, 1987 and Bilgrami *et al.*, 1991) — Appendices I and II. *Lepista macrospora* Rawla & Brar, new to science and *Clitocybe petchii* Pegler and *C. squamulosa* (Pers. ex Fr.) Kummer, new records for India are being described and illustrated in this paper. A key to the identification of all the known species of the two genera so far recorded from India is also provided.

Material and Methods

The collection of the fungi were made from Shimla hills. Their field notes, ecological nich, were noted. The spore print was obtained from the fresh specimens. The material was dried at 60°C. The exicatii were wrapped in paper packets (6 × 4 inches) alongwith naphthelene balls and deposited in PAN (Panjab University herbarium) giving PAN no. on top left, collection no. on top right, name, date of collection, name of locality from where collected, altitute, substrate, name of collector and determinant with date of determination and any other relevant reference on the slip. The dried material was revived in 2% or 8% KOH and stained with ammonical erythrosine (8%) for microscopic study. The other chemical tests were made after Moser (1983) and Singer (1986).

* Botany Department, Panjab University, Chandigarh-160 014

The camera lucida illustrations were drawn at 2,000 magnification and reduced accordingly. The spore measurements are an average of hundred spores. The colour terminology is after Rayner (1970). The Q value of the spores is length divided by breadth.

Clitocybe Kummer,
Fuhr. Pilzk.: 26 (1871).

Key to the Indian Species of *Clitocybe*

1. Stipe excentric 2.5–3.5 cm long, strigose at the base; pileus 2.5–5 cm diameter; spores 4–5 × 2.5–3 µm .... 1. ***C. excentrica***
1. Stipe central 2
2. Epicutis of the pileus containing numerous vesiculose swollen elements. Pileus grey, hygrophanous, 3–5 cm diameter; stipe up to 6 cm long; spores 5.5–6.5 × 3–3.5 µm .... 2. ***C.hydrogramma***
2. Epicutis of the pileus devoid of vesiculose swollen elements 3
3. Pileus always large, up to 14 cm diam; fleshy, convex to applanate, non-hygrophanous, brown gray or buff gray or gray, without greenish tinges, spores ellipsoid or oblong ellipsoid 4
3. Pileus medium to large, usually depressed to infundibuliform 5
4. Spores 4–5 × 2.5–3 µm .... 3. ***C. nebularis***
4. Spores 6–10 × 3.5–5 µm .... 4. ***C. clavipes***
5. Pileus pale buff or fulvous, pinkish buff to chestnut, red brown to umber brownish, neither viscid nor hygrophanous 6
5. Pileus either white or shining white or becoming white and then with some dull grayish blue, grayish-straw or cinnamon apricot to sordid cream or grayish shade; ± hygrophanus from the beginning or in age only, but often viscid 9
6. Spores mostly small to medium sized, usually globose or piriform or lacrymoid; basidiocarps usually odourless 7
6. Spores usually medium sized, often broadly ellipsoid; basidiocarps often with a strong odour 8
7. Spores globose, 3–4 µm, minutely warted .... 5. ***C. flaccida***
7. Spores short ellipsoid to obovate or piriform in face view, often sublacrymoid in side view, smooth .... 6. ***C. squamulosa* var. *squamulosa***
8. Stipe 1–2 cm × 4–6 mm, attenuated downwards often compressed .... 7. ***C. incilis***
8. Stipe 3–5 cm × 0.5–9 mm, equal, cylindrical .... 8. ***C. sinopica***
9. Pileus up to 4 cm diameter 10
9. Pileus up to 5 cm diameter 13
10. Pileus vinaceous gray with blackish centre or apricot yellow, spores oblong to ellipsoid or globose; stipe ± radicating 11
10. Pileus white; spores minute, 4 × 3 µm 12
11. Pileus vinaceous gray with blackish centre; stipe non-radicating; spores ellipsoid or oblong, 4.5–7.5 × 3–4 µm .... 9. ***C. petchii***
11. Pileus apricot yellow all over; stipe radicating; spores globose, 3.5–4.5 × 3.5–4.5 µm .... 10. ***C. mayyanadensis***
12. Pileus 2–4 cm diameter, plano-convex; stipe up to 3 cm long .... 11. ***C. dealbata***

12. Pileus 1–2 cm long diameter, conves depressed; stipe 1 cm long .... 12. *C. pumila*
13. Lamellae interveined; pileus shining white, becoming yellowish with age; stipe 2–3 cm × 4–5 mm; spores 4–5 × 2.5–3 μm, minutely roughened .... 13. *C. ericetorum*
13. Lamellae non interveined 14
14. Pileus livid (bluish gray) when moist, becoming whitish; stipe 3–6 cm × 3–8 mm, grayish then whitish, straight, equal or slightly thickened at the apex; spores ellipsoid, 5–6 × 3 μm .... 14. *C. brumalis*
14. Pileus cream or flesh coloured, becoming shining white; stipe 5–7 cm × 2–3 mm, concolorous, often curved or flexuose, non radicating; spores ovoid, 3–4 × 2–3 μm .... 15. *C.anguistissima*

Clitocybe petchii Pegler, Kew Bull. Addit. Ser. 7:84 (1986).
C. sordida Petch, Ann. Roy. Bot. Gard., Peradeniya 7:279 (1922).

Pileus up to 3 cm diam, plano convex soon depressed to slightly infundibuliform, initially vinaceous gray, gray or blackish brown at the centre, somewhat disrupted into dense, minute, appressed squamules, radially striate; margin at first incurved, whitish. Lamellae decurrent, white to pallid, narrow, somewhat arcuate, moderately crowded with lamellulae of three lengths. Stipe up to 4 cm long, up to 4 mm wide, cylindric, central, fistulose with a subbulbous base; surface greyish buff, minutely striate, squamulose; veil none. Context up to 1 mm thick at the centre, thin, hyaline or pale yellowish, inamyloid, consisting of thin-walled hyphae, 4.5–9 μm diam. Odour faintly farinaceous; taste not distinct. Spore print pale whitish. Spores 4.5–7.4 × 3.4 (6 × 3.4) μm, Q = 1.7, broadly ellipsoid or oblong, hyaline, inamyloid, smooth, thin-walled, with few contents. Basidia 24.5–32 × 5.5–7 μm, narrowly clavate, 4-sterigmatus; sterigmata (4–5 × 1–1.5 μm). Cheilocystidia poorely differentiated, few, 19–25 × 5.5–6 μm, lageniform to clavate or ventricose rostrate, subhyaline, thin-walled. Pleurocystidia not differentiated. Hymenophoral trama more or less regular, inamyloid, consisting of slightly thick-walled, hyaline hyphae, 4.5–9 μm diam. Subhymenial layer well developed, 12–15 μm wide, densely interwoven, inamyloid. Peleipellis a poorely differentiated cutis, consisting of somewhat subparallel, thin-walled, pale yellowish or hyaline hyphae, 3.5–12.5 μm diam, a few hyphae tending to form semierect hyphous elements but never forming a true trichodermium. Pigments intra-cellular, pale yellowish. All hyphae are clamped (Fig. 1).

Habit, Habitat and Distribution

Solitary or subcaespitose (and then the basidomes are connate), humicolous on decaying vegetation, in moist deodar forest, Mashobra,

Shimla (HP) 1700–2500 m. A monsoon species requiring moderate light, moisture and temperature; locally common.

Material Examined

India, Himachal Pradesh Hills : Shimla, Mashobra, August 6, 1992 (PAN 103079 = 40/92); leg. K.P.S. Brar.

The specimen in its macroscopic and microscopic details, matches well with the description of *Clitocybe petchii* Pegler (Pegler, 1986), except having cheilocystidia.

It is a maiden record for India.

Clitocybe squamulosa (Pers. ex Fr.) Kummer var. ***squamulosa*** Bigelow, Lloydia 31 (1): 51–53 (1968)

Agaricus squamulosa Fries, Syst. Myc. 1: 82 (1821).
Clitocybe sinopicoides Peck, N.Y. State Mus. Bull. 157: 80 (1912).
Clitocybe subsquamata Murrill, Mycologia 7: 274 (1915).

Pileus up to 7 cm diam, plane or convex-depressed, becoming broadly infundibuliform, sometimes umbilicate, light ochraceous to umber at the centre with ochraceous buff periphery, plae in age or in dry weather, glabrous when moist, otherwise dry, subhygrophanous, often squamulose about centre; margin incurved at first, at times ridged, striate, entire or at times undulate to crenate to incised or lobed. Lamellae moderately decurrent to rather long decurrent, light buff, close to subdistant, narrow to moderately broad, forked and/or intervenose at times, with lamellulae of three lengths; edge even except in age. Stipe up to 4.5 cm long, up to 8 mm wide, central, cylindrical, fistulose to hollow in age, fibrous tough equal or somewhat swollen below, often with tomentose or mycelial base; surface concolorous with the pileus or somewhat pale luteus, glabrous or fibrillose, striate with white appressed fibrils. Veil none. Context up to 2 mm thick at the centre, pliant or flabby when wet, firm or brittle when dried, watery buff to pallid, inamyloid, consisting of loosely interwoven, thin-walled, hyaline hyphae, 4.5–10.5 (15) μm diam. Taste not distinct; odour fungoid to somewhat farinaceous. Spore print whitish. Spores 5.5–9 × 3.5–5.5 (7 × 4.5) μm, Q = 1.6, short ellipsoid to obovate or piriform in face view, often sublacrymoid in side view, hyaline, inamyloid, acyanophilous, smooth, thin-walled, with very few contents. Basidia 23–33 × 5–7 μm, clavate, hyaline, thin-walled, 4-sterigmatus; sterigma (6–7 × 1–1.5 μm) with slightly bent apices. Basidioles numerous, filamentous. Cheilocystidia poorely differentiated, few,

20–29 × 6–8 μm, subcylindric to calvate, hyaline, thin-walled. Pleurocystidia not differentiated. Hymenophoral trama interwoven to subregular, hyaline, inamyloid, consisting of thin-walled, pale yellowish or hyaline hyphae, 6–12 μm diam. Subhymenial layer well developed, up to 15 μm wide, hyaline, appearing pseudoparenchymatous, inamyloid. Pileipellis a poorely differentiated cutis, consisting of loosely interwoven, pigment encrusted, slightly thick-walled, cylindrical hyphae, 4.5–7 μm diam, with protruding end cells at times. Pigments encrusted, brownish to dingy yellowish in KOH. All hyphae clamped (Figure 2).

Habit, Habitat and Distribution

Solitary, scattered or rarely subcespitose, humicolous amongst decaying conifer needles, in moist deodar forest, Mashobra, Shimla (HP) 1700–2500 m. A monsoon species requiring shade, moderate moisture, and temperature; locally uncommon.

Material Examined

India, Himachal Pradesh Hills : Shimla, Mashobra, August 5, 1992 (PAN 103080 = 17/92); leg. K.P.S. Brar.

According to Singer (1986), the specimen comes in section *Clitocybe* (Subgenus *Clitocybe*) because of fleshy pileus, which is depressed to infundibuliform, ochraceous to umber brown, subhygrophanus; dry squamulose, visid only when moist; deeply decurrent lamellae; stipe without rhizoid mycelial fibrils but with poorely differentiated basal mycelium or tomentum; whitish spore print; ovoid to piriform, acynophilic spores. In the section *Clitocybe,* it comes in sub section *Clitocybe* because of ovoid to piriform spores and white basal tomentum. In the subsection *Clitocybe* it resembles closely the description of *C. squamulosa* (Fr.) Kummer, var *squamulosa* as described by Bigelow (1968) in all its essential details except that if differs from it in having few cheilocystidia. It is a new report for India.

Lepista (Fr.) W.G. Smith,
Clavis Agar 26 (1870).

Key to the Indian Species of *Lepista*

1. Basidiocarp bluish-violet or lilac, pale blue or purplish; spore print pinkish .... 2. **Sect.** ***Spongiosa***
1. Basidiocarp without violet or lilac or purplish pigments 4

2. Pileus purple to violaceous, drying to pale luteus, non-hygrophanous, up to 3 cm diam, stipe up to 3.7 cm long, 0.25 cm thick, sordid pink to lilac, with a bulbous base; spores 4.8–7.3 × 3.5–4.5 μm .... 1. ***L. nudoidea***
2. Pileus bluish lilac to violet or grayish lilac to brownish violet; spores 5–7 × 3–4.5 μm 3
3. Pileus up to 10 cm diameter, bluish lilac to violet; drying to reddish brown, hygrophanous, stipe without a bulbous base; spores 5–7 × 3.5–4.5 μm .... 2. ***L. nuda***
3. Pileus 2–7 cm diameter, grayish lilac to brownish violet, drying to deep brown, non hygrophanous, stipe with a bulbous base; spores 6–7 × 3–4 μm .... 3. ***L. sordida***
4. Spore print pinkish cream to salmon; Basidio carps clitocyboid to tricholomatoid, often large, stipe white or light pinkish cinnamon 5
4. Spore print white or most commonly yellow or pinkish buff, stipe concolorous with pileus .... 6. **Sect. *Gilva***
5. Pileus up to 12 cm diam; brown; stipe white; spores 5.7–7.5 × 3.5–5.5 μm .... 4. ***L. glabella***
5. Pileus up to 15 cm diam; pale pinkish buff; stipe whitish to light pinkish cinnamon; spores 3.5–4.5 × 2.5–3.5 μm .... 5. ***L. densifolia*** (= *L. subaqualis*)
6. Spores large, 6.5–10 × 3.5–6 μm 7
6. Spores small, 4–7 × 3–4 μm 8
7. Pileus up to 6 cm diam, convex depressed, becoming planoconvex to infundibuliform, egg yellow all over, smooth; spore print yellowish cream; spores 6.5–9.5 × 3.5–5.5 μm .... 6. ***L. macrospora sp. nov***
7. Pileus upto 2.5 cm diam, convex to umbonate, becoming plane, creamish to reddish brown, darker on umbo; scaly, scales darker; spore print pinkish buff; spores 7–10 × 5–6 μm .... 7. ***L. kamati***
8. Basidiocarps caespitose; pileus buff, hazel on drying; surface smooth and glabrous; spores 5.5–7.0 × 3.5–4.0 μm, ellipsoid .... 8. ***L. indica***
8. Basidiocarps solitary; pileus buff-yellow; surface with numesous, minute fuscous brown appressed squamules, concentrated more at the disc; spores 4–5.6 × 3–4 μm, subglobose to broadly ellipsoid .... 9. ***L. lentiginosa***

Lepsita macrospora Rawla & Brar, *sp. nov.*
Etym: Gk: macro = large; large spored.

Basidiocarp magnus ad 6 cm × 6 cm, clitocybioideus, luteus, pileus aviscidus, nudus. Sporae 6.5–9.5 × 3.5–5.5 μm ornatae, magnae, minuto punctatae e quamo-hilare plage. Lamellae luteae decurrenteae. Cystidia nulla. Pellicula cuticulae ex hyphus repentibus et subparallelae oder intricatae.

Habitat

In humo solitar, in folioputridum *Cedrus deodara* Loud, silvis Shimla, August 6, 1992, 103081, India, Holotybes in PAN conservatus, legit Brar.

Pileus up to 6 cm diam, convex depressed at first becoming plano-despressed then infundibuliform; with involute margin, egg-yellow, evenly colored, smooth, glabrous and naked, neither hygrophanous nor visid; margin entire, involute, concolorous. Lamellae decurrent, concolorous with pileus but much lighter (light yellowish gilvous), arcuate, moderately crowded with lamellulae of three lengths; edge concolorous, dentate. Stipe up to 6 cm long, up to 8 μm, central, cylindrical, fistulose or hollow with a subradicating base; surface concolorous with the pileus, subglabrous, fibrillose; fibrilis appressed. Viel none. Context up to 2 mm thick at the centre, watery ochraceous, firm, inamyloid, consisting of hyaline, thin-walled hyphae, 6–12 μm diam. Smell and taste none. Spore print yellowish cream. Spores 6.5–9.5 × 3.5–5.5 (7.7 × 4.4) μm, Q = 1.75, ovoid to broadly ellipsoid, sub-hyaline, inamyloid, evenly cyanophilus, slightly thin-walled, minutely punctate roughened in outline, a few with a smooth, suprahilar spot or plage. Basidia 31–38 × 7–9 μm, clavate, thin-walled, 4-sterigmatus; sterigmata 1.5–2 × 1–1.5 μm. Pleurocystidia and cheilocystidia not differentiated. Hymenophoral trama regular, but consisting of subparallel to parallel, subhyaline, inamyloid, slightly thick-walled hyphae, 3–6 μm diam, generally axillary arranged, only here and there in the central strand of the trama but more irregularly interwoven towards the subhymenium. Subhymenial layer poorely developed, up to 9 μm wide, interwoven, inamyloid. Pileipellis an undifferentiated cutis, consisting of appressed subparallel and in places interwoven, thin-walled and non incrusted, hyaline-yellow, filamentous hyphae, 4–11.0 μm wide. Pigments scanty if present granular, in the form of yellowish granulations. All hyphae are clamped (Fig. 3).

Habit, Habitat and Distribution

Solitary, humicolous amongst the decaying needles of *Cedrus deodara* Loud., in moist deodar forest, Mashobra, Shimla (HP) 1700–2500 m. A monsoon species requiring moderate temperature, light and moisture; locally uncommon.

Material Examined

India, Himachal Pradesh Hills. Shimla, Mashobra, August 6, 1992 (PAN 103081 = 36/92- Holotype); leg. K.P.S. Brar.

According to Singer (1986) the species is keyed out to Section *Gilva* because of egg-yellow carpophores; clitocyboid habit; yellow spore print; minutely punctate-rough spores, with suprahilar plage. In the section

Gilva, it does not resemble any of the previously known species except *A. ameliae* (Arcangeli) Sing. & Clemencon, to which it resembles in the egg yellow, smooth, non-visid, naked pileus with sharply depressed centre and involute margin, concolorous stipe and light yellowish gilvous decurrent lamellae; yellow spore print; absence of cystidia and poorely differentiated epicutis consisting of appressed, subparallel hyphae which at places, are somewhat interwoven. However, it differs outrightly from *L. ameliae* (Arcangeli) Singer & Clemencon and all other previously known species in the section *Gilva*, in having large spores (6.5–9.5 × 3.5–5.5 μm); large basidia (31–38 × 7–9 μm) and slightly large basidiocarps (pileus reaching up to 6 cm diam and stipe up to 6 cm long). Therefore, on this account it is being treated as a new species with following distinguishing characteristics.

Basidiocarps large up to 6 cm diam, and up to 6 cm high, egg yellow; Basidiospores large, 6.5–9.5 × 3.5–5.5 μm, with suprahilar plage; Basidia large 31–38 × 7–9 μm.

REFERENCES

Bilgrami, K.S., Jamaluddin, S. and Rizwi, M.A., 1991. *Fungi of India*. Lists and References. Today and Tomorrow's Printers and Publishers, New Delhi.

Bigelo, H.E., 1968. The genus *Clitocybe* in North America II Section *infundibuliformes*. Lloydia. **31** (1) 43–62.

Manjula, B., 1983. A revised list of agaricoid and boletoid basidiomycetes from India and Nepal. Proc. Indian Acad. Sci. (Plant Sci.) **92** 81–215

Moser, 1983. *Die Rohrlinge and Blatterpilze Kleine Kryptogamenflore* II b/2. 5. Aufl Fischer Stuttgart 533 p.

Pegler, D.N., 1986. Agaric flora of Sri Lanka, Kew Bull Add. Ser. **12** 1–519.

Rawla, G.S. and Arya, S., 1987. Some edible and poisonous agarics new to India, in T.N. Koul and B.M. Kapoor (Eds.) *Indian Mushroom Science*: II *Science and cultivation technology of edible fungi*. Regional Research Laboratory CSIR, Jammu Tawi, pp. 375–382.

Rayner, R.W., 1970. *A mycological colour chart*; CMI, Kew, Surrey, London.

Sathe, A.V. and Daniel, J., 1980. *Agaricales (Mushrooms) of South West India* p 1. *Agaricales (Mushrooms) of Kerala state*. MACS Res. Inst. Monogr No. **1** 75–108.

Sathe, A.V. and Deshpande, 1980. *Agaricales (Mushrooms) of South West India* p I, *Agaricales (Mushrooms) of Maharashtra state*. MACS Res. Inst. Monogr No. **1**: 9–42.

Sathe, A.V. and Kulkarni, S.M., 1987. A check list of wild edible mushrooms from South West India in T.N. Koul and B.M. Kapoor (Eds.) *Indian Mushroom Science* : II *Science and cultivation technology of edible fungi*. Regional Research Laboratory, CSIR Jammu Tawi, pp. 411–413.

Singer, R., 1986. *The Agaricales in Modern Taxonomy* 4th Ed. Vaduz J. Cramer Germany.

Verma, R.N., Singh, G.B. and Bilgrami, K.S., 1987. Fleshy fungal flora of the NEH India-I, Manipur and Meghalaya in T.N. Kaul and B.M. Kapoor (Eds.). *Indian Mushroom Science* : II *Science and cultivation technology of edible fungi*. Regional Research Laboratory, CSIR, Jammu Tawi, pp. 414–421.

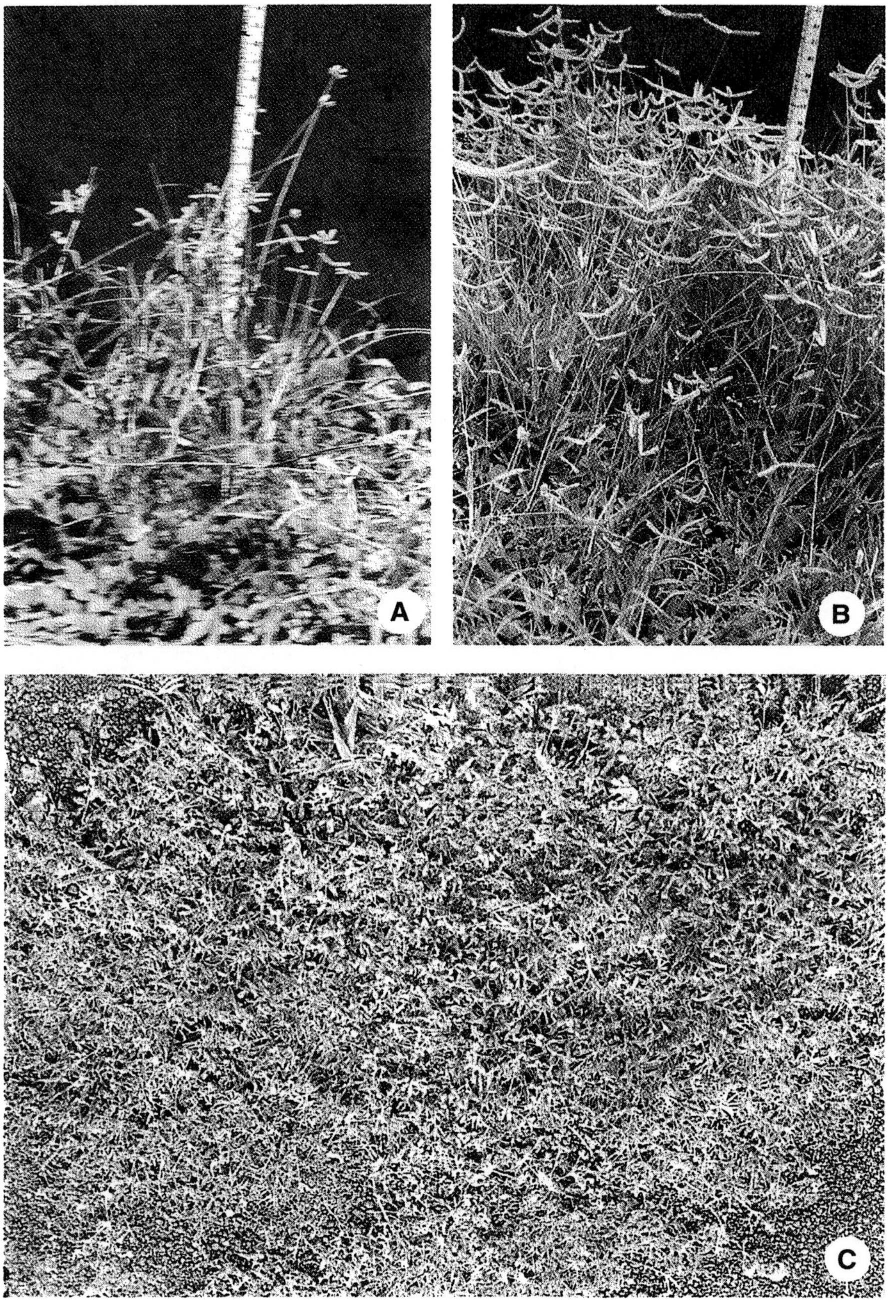

PLATE 1

Ecotypes of *D. aegyptium* — A: C_4 form; B: L_2 form; C: C_5 form

Appendix - I

List of species of *Clitocybe* from India (Bilgrami *et al.*, 1991)

1. *C. angustissima* (Lasch.) Kummer
2. *C. brumalis* (Fr.) Quél.
3. *C. clavipes* (Fr.) Kummer
4. *C. ericetorum* (Bull.) Fr.
5. *C. excentrica* Peck.
6. *C. flaccida* (Sow. ex Fr.) Kummer
7. *C. inclinis* (Fr.) Quél.

C. incongrua Berk.	transferred to *Leptonia incongrua* (Manjula, 1983)
C. laccata (Scop.) Sacc.	transferred to *Laccaria laccata* (Manjula, 1983)

8. *C. pumila* Massee.
9. *C. sinopica* Christ

C. stipitaria Fr.	transferred to *Crinipellis stipitaria* (Singer, 1986)
C. tabescens (Scop.) Brés.	transferred to *Armillaria tabescens* (Manjula, 1983)
Clitocybe sp.	doubtful and excluded

Supplementary List

10. *C. dealbata* (Sow. ex Fr.) Kummer (Manjula, 1983)
11. *C. hydrogramma* (Bull. ex Fr.) Kummer (Manjula, 1983)
12. *C. nebularis* (Batsch ex Fr.) Quél. (Rawla and Arya, 1987).
13. *C. mayyanadensis* Sathe & Kulkarni (Sathe and Kulkarni, 1980).

Appendix - II

List of species of *Lepista* from India (Bilgrami *et al.*, 1991)

1. *L. glabella* (Speg.) Sing.
2. *L. kamati* Sathe & Sasangam.
3. *L. subsqualis* (Bri.) Singer transferred to *Lepista densifolia* (Singer, 1986)

Supplementary List

4. *L. sordida* (Fr.) Singer
   (Manjula, 1983)
5. *L. nudoidea* Sathe & Deshpandae
   (Sathe and Deshpandae, 1980)
6. *L. nuda* (Bull. ex Fr.) Cooke
   (Sathe and Kulkarni, 1987).
7. *L. indica* Sathe & Daniel
   (Sathe and Daniel, 1980).
8. *L. lentiginosa* (Fr.) Pegler
   (Verma *et al.*, 1987).

CHAPTER 2

THE STUDIES ON GENETIC VARIABILITY IN CROWFOOT GRASS (*DACTYLOCTENIUM AEGYPTIUM* (L.) BEAUV.) POACEAE I. ECOTYPES

T.N. Mary and M. Vijaya Koteswari***

Introduction

All members of a local population share a single gene pool and any one of them has equal probability of mating with any other in producing offspring (Mayor, 1970). A common mode of speciation in cross-fertilizing plants involves gradual divergence of two or more population systems. Hence, the study of natural populations has gained considerable importance in recent years and they are being recognized as the fundamental units of a species.

The studies on the genetical basis of ecological differentiation within populations of species were aptly described as genecology (Turesson, 1922a). The existence of habitat-associated races or ecotypes in plants is widespread (Turesson, 1922b; Clausen *et al.*, 1940, 1948; Heslop-Harrison, 1964 and Langlet, 1971). Nevertheless, the forces leading to habitat specificity have most often remained poorly understood. Genetic differentiation between adjacent populations has been recognized as a balance between gene flow and selection (Jain and

* Professor, Department of Botany, Nagarjuna University, Nagarjunanagar-522 510, A.P.

** Lecturer, Hindu College, Guntur-522 001, A.P.

Bradshaw, 1966) and ecotypic differentiation over short distances may be similarly controlled. The morphological variability, chromosome number, and phenology of different populations belonging to crowfoot grass *Dactyloctenium aegyptium* (L.) Beauv has been discussed (Mary and Vijaya Koteswari, 2003). The present article aims at analysing the causes of ecotypic differentiation (Fig. 1) among these natural populations of the crowfoot grass, *Dactyloctenium aegyptium* (L.) Beauv from in and around the grasslands of the Nagarjuna University campus.

Results

D. aegyptium is an annual, appearing in rainy and winter seasons. A critical study of its populations revealed the occurrence of eight morphologically distinct forms recognized by their size, stature, colour, spike morphology, number of tillers, spikes/inflorescence etc. The plants are basically divided into two main groups based upon the spike morphology–compact (C_1, C_2, C_3, C_4 and C_5) and lax types (L_1, L_2 and L_3).

Morphological data collected for all the eight forms (C_1, C_2, C_3, C_4 and C_5) and (L_1, L_2 and L_3) are summarized in Table 1. The 't' values of the same are represented in Table 2. The above data bring out the quantitative comparison of the morphological features of the forms in question. 't' test, when applied between the populations (Table 2), showed significant differences for root, shoot, leaf, spike, etc. The character-association analysis for the eight morphological parameters studied now can be differentiated under five well-defined complexes (Table 3, Plate 1) according to Anderson (1957).

Complex A

C_1 (2n=46) comes under this complex, consisting of the plants showing less number of leaves and shorter shoot length. As regards the other characters, the plants showed minimum values. The range of variability in the characters considered is much less.

Complex B

This group includes the prostrate, compact form, C_5 (Plate 1, 2n=46) with many leaves, but short shoot length. The values for the other morphological characters are intermediate.

Complex C

This includes lax, prostrate form, L_3 (2n=38) with shorter shoot length than in the C_5 complex. The values for other morphological characters are intermediate as in Complex B.

Complex D

C_4 (2n=46) comes under this group with erect, long shoots and with more number of leaves. The values for other characters are maximum. The range of variability is more in comparison to Complex A.

Complex E

C_2, C_3 (2n=46), L_1 and L_2 (2n=40) (Plate 1) fall under this complex because of the similarities in shoot length and number of leaves. Within this complex, the different forms are nearly isolated as evident from the positions they occupy in the scattered diagram. The range of variability for the characters considered is medium. C_2 and C_3 did not show perceptible values for length of rachilla, whereas the lax forms, L_1 and L_2 showed maximum values. Overall, it can be said that the populations are heterozygous.

An examination of the tabulated data reveals that there are significant differences among the forms for the length of shoot, leaf, spike, internode, and rachilla, number of flowers/spike, spikes/inflorescence etc. The same is evident from the Anderson's scattered diagram also (Fig. 1).

Culture Experiments

(For methods followed see Mary and Vijaya Koteswari, 2003).

Common-garden experiment

Under uniform environmental conditions, C_4, C_5 and L_2 alone retained their respective characters. However, C_1, C_2 and C_3 appeared similar to C_4 as did L_1 and L_3 to L_2 (Table 4, Plate 1: a–c). It can be said that environmental factors on intermediate plants, i.e., C_1, C_2, C_3, L_1 and L_3 were less effective when the characters like the number and nature of spike, length of rachilla were considered, while some of the other characters like length of root, shoot, number of tillers and leaves appeared to be very much under environmental control.

Table 1. Mean and standard deviation values of certain characters in the eight forms of *D. aegyptium*

Form	Root length (cm)	Shoot length (cm)	Number of tillers	Number of leaves	Leaf length (cm)	Leaf breadh (cm)	Flag leaf length (cm)	Flag leaf breadth (cm)	Number of inflore-scences	Number of spikes	Length of spikes (cm)	Number of spike-lets	Length of peduncle (cm)	Length of 2nd inter-node (cm)	Length of rachilla (cm)
C_1	5.17 ±0.23	7.17 ±0.31	1.00 ±0.00	6.07 ±0.19	2.01 ±0.11	0.22 ±0.01	0.79 ±0.07	0.11 ±0.03	1.00 ±0.00	1.00 ±0.00	0.56 ±0.03	8.10 ±0.20	2.39 ±0.33	1.56 ±0.06	0.10 ±0.00
C_2	7.92 ±0.34	29.42 ±0.97	1.00 ±0.00	7.90 ±0.31	5.58 ±0.38	0.38 ±0.01	2.04 ±0.18	0.26 ±0.01	1.00 ±0.00	2.00 ±0.00	1.61 ±0.06	31.80 ±1.42	9.31 ±0.44	5.58 ±0.20	0.10 ±0.00
C_3	11.20 ±0.41	32.83 ±0.74	3.83 ±0.23	32.73 ±1.81	4.93 ±0.21	0.44 ±0.01	1.97 ±0.13	0.26 ±0.01	4.07 ±0.35	3.47 ±0.09	1.65 ±0.06	36.83 ±1.26	10.21 ±0.51	4.12 ±0.17	0.10 ±0.00
C_4	12.40 ±0.51	74.70 ±2.60	7.87 ±0.62	75.60 ±6.04	12.77 ±0.67	0.67 ±0.03	5.58 ±0.30	0.40 ±0.01	8.40 ±0.69	4.10 ±0.06	2.76 ±0.04	61.00 ±1.53	19.46 ±0.74	11.76 ±0.33	0.10 ±0.00
C_5	15.63 ±0.60	26.40 ±1.06	6.20 ±0.28	120.93 ±6.01	4.67 ±0.18	0.47 ±0.02	1.53 ±0.09	0.32 ±0.01	10.10 ±1.04	4.00 ±0.00	1.56 ±0.07	30.37 ±0.90	5.87 ±0.37	4.23 ±0.22	0.10 ±0.00
L_1	5.17 ±0.19	21.75 ±0.67	1.47 ±0.11	11.13 ±0.89	4.19 ±0.14	0.36 ±0.01	2.80 ±0.11	2.27 ±0.01	1.33 ±0.11	1.00 ±0.00	2.10 ±0.08	34.70 ±1.35	12.67 ±0.42	3.05 ±0.15	0.38 ±0.01
L_2	9.20 ±0.44	39.23 ±0.95	2.73 ±0.23	24.40 ±1.82	6.41 ±0.25	0.48 ±0.02	3.91 ±0.12	2.37 ±0.02	3.70 ±0.25	2.27 ±0.08	3.38 ±0.07	51.63 ±1.31	14.78 ±0.50	4.55 ±0.30	0.40 ±0.00
L_3	8.15 ±0.41	12.15 ±0.71	10.23 ±0.56	113.33 ±6.30	3.08 ±0.12	0.42 ±0.01	2.25 ±0.08	0.35 ±0.01	22.77 ±1.11	2.00 ±0.00	1.87 ±0.06	12.23 ±1.18	4.65 ±0.18	2.50 ±0.12	±0

Table 2. 't' test values between the forms for different characters
Matrix of 't' Values

(a) Root length

Form	C_1	C_2	C_3	C_4	C_5	L_1	L_2
C_2	21.916						
C_3	32.104	2.810					
C_4	26.271	16.592	15.762				
C_5	17.442	−2.104	−4.988	−17.481			
L_1	19.777	−6.516	−11.131	−20.068	−3.712		
L_2	32.215	7.254	5.336	−13.030	9.037	15.078	
L_3	6.432	−14.384	−20.204	−23.611	−11.175	−9.829	−22.881

(b) Shoot ength

Form	C_1	C_2	C_3	C_4	C_5	L_1	L_2
C_2	6.181						
C_3	8.931	5.497					
C_4	10.013	7.367	1.694				
C_5	12.539	11.282	5.727	4.130			
L_1	3.525	−7.081	−11.359	−13.299	−16.691		
L_2	7.172	2.298	−3.004	−4.734	−8.641	8.279	
L_3	6.233	0.438	−4.736	−6.493	−10.319	6.527	−1.726

(c) Number of tillers

Form	C_1	C_2	C_3	C_4	C_5	L_1	L_2
C_2	9.999						
C_3	12.575	12.575					
C_4	11.025	11.025	6.089				
C_5	18.491	18.491	6.568	−2.439			
L_1	4.065	4.065	−9.359	−10.106	−15.583		
L_2	7.240	7.240	−3.346	−7.693	−9.387	4.771	
L_3	16.533	16.533	10.627	2.829	6.450	15.376	12.343

(d) Number of leaves

Form	C_1	C_2	C_3	C_4	C_5	L_1	L_2
C_2	5.097						
C_3	1.132	0.820					
C_4	11.491	11.179	7.448				
C_5	19.097	18.778	12.862	5.321			
L_1	5.564	3.437	−0.269	−10.545	−18.067		
L_2	9.978	8.907	1.894	−8.102	−15.363	6.527	
L_3	17.019	16.716	11.669	4.326	−0.873	16.063	13.558

(e) Leaf length

Form	C_1	C_2	C_3	C_4	C_5	L_1	L_2
C_2	9.173						
C_3	12.493	–1.515					
C_4	15.822	9.339	11.139				
C_5	12.894	–2.212	–0.975	–11.664			
L_1	12.333	–3.468	–2.933	–12.491	–2.089		
L_2	16.121	1.831	4.519	–8.864	5.671	7.676	
L_3	6.071	–7.984	–57.254	–15.150	–14.925	–7.422	–23.071

(f) Leaf breadth

Form	C_1	C_2	C_3	C_4	C_5	L_1	L_2
C_2	10.521						
C_3	11.906	3.519					
C_4	14.083	9.146	6.722				
C_5	12.374	4.614	1.190	–5.773			
L_1	9.343	–1.712	–4.944	–9.975	–5.940		
L_2	13.833	5.663	1.877	–5.495	0.589	7.120	
L_3	10.446	2.169	–1.086	–7.369	–2.196	3.519	–2.919

(g) Flag leaf length

Form	C_1	C_2	C_3	C_4	C_5	L_1	L_2
C_2	6.456						
C_3	7.774	–0.317					
C_4	15.662	10.258	11.153				
C_5	6.218	–2.512	–2.699	–13.035			
L_1	15.173	3.664	4.889	–8.811	8.864		
L_2	21.623	8.670	10.811	–5.212	15.440	6.761	
L_3	12.787	1.082	1.809	–10.799	5.682	–3.973	–11.082

(h) Flag leaf breadth

Form	C_1	C_2	C_3	C_4	C_5	L_1	L_2
C_2	9.742						
C_3	12.155	0.198					
C_4	18.488	7.262	7.880				
C_5	0.174	–2.275	–2.357	–4.652			
L_1	14.019	1.019	0.952	–7.300	2.593		
L_2	15.211	5.405	5.757	–1.442	4.116	5.154	
L_3	21.409	5.967	6.743	–2.586	3.978	6.026	–0.742

(i) Number of inflorescences

Form	C_1	C_2	C_3	C_4	C_5	L_1	L_2
C_2	9.999						
C_3	8.875	8.875					
C_4	10.686	10.686	5.599				
C_5	8.714	8.714	5.485	1.357			
L_1	3.010	3.010	−7.533	−10.077	−8.348		
L_2	11.012	11.012	−0.865	−6.398	−5.967	8.797	
L_3	19.524	19.524	16.021	10.947	8.292	19.131	16.703

(j) Number of spikes

Form	C_1	C_2	C_3	C_4	C_5	L_1	L_2
C_2	9.999						
C_3	26.626	15.832					
C_4	42.177	28.571	5.356				
C_5	9.999	9.999	5.757	−1.361			
L_1	9.999	9.999	−26.626	42.177	9.999		
L_2	15.425	3.247	−9.693	−16.635	−21.108	15.425	
L_3	9.999	9.999	−15.832	−28.571	9.999	9.999	−3.247

(k) Length of spike

Form	C_1	C_2	C_3	C_4	C_5	L_1	L_2
C_2	14.661						
C_3	15.376	0.438					
C_4	38.458	14.209	13.827				
C_5	13.571	−0.535	−0.969	−14.494			
L_1	17.504	4.663	4.303	-6.895	5.069		
L_2	36.133	18.282	17.970	7.162	18.500	11.692	
L_3	18.655	2.835	2.408	−11.237	3.325	−2.244	−15.826

(l) Number of flowers

Form	C_1	C_2	C_3	C_4	C_5	L_1	L_2
C_2	14.838						
C_3	19.693	2.623					
C_4	31.249	13.937	12.139				
C_5	19.317	−0.870	−4.153	−17.208			
L_1	17.288	1.458	−1.148	−12.839	2.659		
L_2	29.160	10.233	8.113	−4.442	13.362	8.975	
L_3	17.444	0.216	−2.655	14.862	1.256	−1.372	−10.999

(m) Length of peduncle

Form	C_1	C_2	C_3	C_4	C_5	L_1	L_2
C_2	14.964						
C_3	12.824	1.220					
C_4	22.809	11.833	9.779				
C_5	8.852	-6.000	-6.225	-16.537			
L_1	23.439	5.527	3.377	-8.045	12.249		
L_2	22.790	7.424	5.222	-5.815	13.507	2.476	
L_3	9.919	-9.771	-8.959	-19.554	-2.969	-17.662	-18.009

(n) Length of II internode

Form	C_1	C_2	C_3	C_4	C_5	L_1	L_2
C_2	19.281						
C_3	13.843	-5.535					
C_4	30.224	15.984	20.411				
C_5	11.543	-4.502	0.412	-18.831			
L_1	8.975	-10.049	-4.588	-23.816	-4.362		
L_2	9.628	-2.831	1.236	-16.020	0.839	4.395	
L_3	6.975	-13.256	-7.651	-26.256	-6.840	-2.817	-6.264

(o) Length of rachilla

Form	C_1	C_2	C_3	C_4	C_5	L_1	L_2
C_2	9.999						
C_3	9.999	9.999					
C_4	9.999	9.999	9.999				
C_5	9.999	9.999	9.999	9.999			
L_1	35.226	22.494	35.226	35.226	35.226		
L_2	9.999	9.999	9.999	9.999	9.999	2.971	
L_3	9.999	9.999	9.999	9.999	9.999	-22.494	9.999

Table value of 't' at 58° of freedom

level	**'t' value**
.05	2.003
.01	2.664
.001	3.469

*9.999 should be considered as 0.

Table 3. Class intervals of the morphological analysis of introgression between the forms of *D. aegyptium*

S.No.	Morphological character	Index values		
		0	1	2
1.	Number of tillers	O 1.0–6.0	O 7.0–14.0	O 15.0–24.0
2.	Length of rachilla	O 0.0–0.1	O 0.2–0.3	O 0.4–0.5
3.	Leaf length	O 1.0–6.0	O 7.0–12.0	O 13.0–19.0
4.	Leaf breadth	O 0.1–0.2	O 0.3–0.4	O 0.5–0.9
5.	Length of spike	O 0.3–1.5	O 1.6–2.8	O 2.9–4.0
6.	Number of spikelets/spike	O 5.0–25.0	O 26.0–45.0	O 46.0–65.0

Reciprocal transplant experiments with tillers

All the transplants of C_4, C_5 and L_2 could establish themselves successfully. About one month later, the transplanted individuals developed tillers and attained full growth. The transplanted plants were distinguished from the indigenous plants, only by their labels. The different forms transplanted at their reciprocal sites could be recognized also by their distinct morphology. No differential survival of plants from different origins could be observed. In all the above three forms, the phenotypic differentiation was continued. The forms showed quicker and higher overall performance at home-sites than at away-sites. Highly significant variation between genets within populations could be detected in several characters. Their differences in mean values were less between the home sites than away-sites for the measured characters viz., root and shoot lengths, and number of tillers, leaves, inflorescences and flowers/spike (Table 5).

Discussion

A striking degree of polymorphicity was observed in the natural populations of the crowfoot grass *D. aegyptium* which existed in an open community and was subject to diverse ecological factors, competition, biotic pressure, etc.

The occurrence of all the forms of *D. aegyptium* as observed on the University campus is attributable to the availability of congenial habitat niches. Though all these forms were capable of growing on the campus in alkaline soils with pH range of 7.6–8.1, each form had confined itself to its natural niche e.g., the prevalence of C_4 in the areas of high moisture content, C_5 in the drier habitats, etc. and L_2 to sandy soils. Spatially, these natural niches were not far apart from one another but only few feet away from one another.

The regular occurrence of the prostrate form (C_5) in the disturbed, and dry habitats, and the erect form (C_4) in the protected and moist areas substantiates the importance of the moisture content, and biotic factors in the distribution of these forms. Such findings are in conformity with the observations made by earlier workers in other species e.g., Jain, 1974; Jameson, 1963; Hanson and Churchill, 1961. Similarly the differences in these two forms may be due to the deviations in their genetic constitution with respect to adaptiveness, reproductive fitness and persistence, which are unique features for each group of species, as has been emphasized by Harper (1970). There are wide discontinuous variations between C_4 and C_5. A low degree of gene flow between them might have been responsible for the observed variation. The

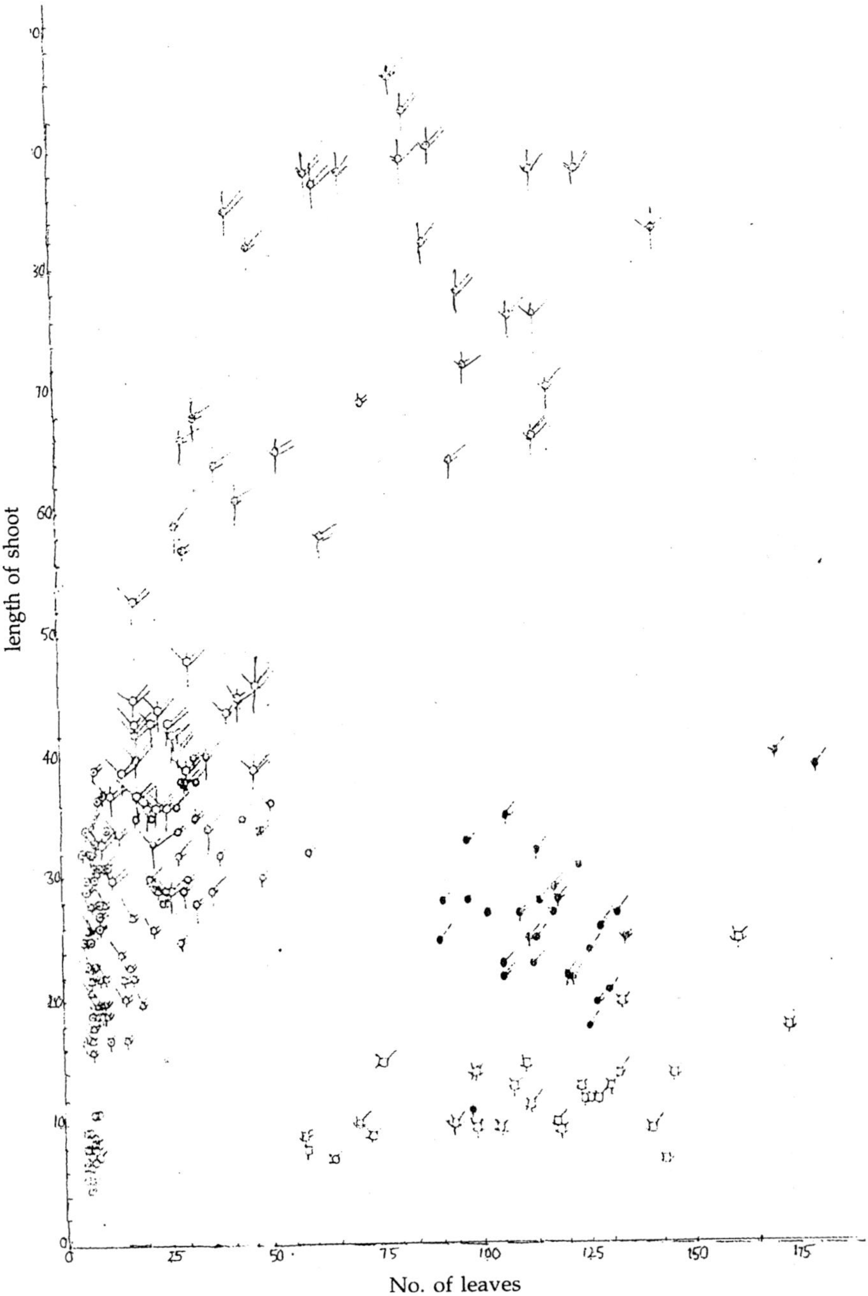

Figure 1. Anderson's graph showing character-association differentiated into five well-defined complexes.

Table 4. Mean and Standard Deviation of certain characters of all the forms of *D. aegyptium* grown under uniform garden conditions

Character	C_1	C_2	C_3	C_4	C_5	L_1	L_2	L_3
Root length (cm)	11.50±0.52	11.30±0.61	10.90±0.61	13.20±0.48	14.00±0.62	9.40±0.40	9.80±0.42	10.10±0.38
Shoot length (cm)	52.80±6.30	53.70±5.20	61.10±7.00	71.90±4.10	32.30±1.08	35.70±0.80	41.20±1.10	33.30±0.70
Number of tillers	5.20±0.20	4.30±0.43	5.10±0.21	6.20±0.60	5.80±0.41	4.70±0.29	6.10±0.33	8.30±0.41
Number of leaves	54.80±4.30	63.30±5.20	69.70±6.80	80.20±7.20	143.70±6.90	30.10±4.20	34.30±6.10	112.20±5.20
Number of Inflorescences	6.20±2.10	7.10±1.80	6.40±1.90	7.40±2.00	12.10±2.20	4.20±1.10	4.40±1.60	6.10±2.10
Number of spikelets/spike	42.40±0.90	41.30±0.80	42.00±0.90	48.80±2.20	36.80±1.80	44.00±1.30	48.00±1.40	44.00±1.23

localization of such forms in different ecological conditions followed by adaptive differences that were maintained by natural selection might have led to the evolution of *D. aegyptium* complex.

A good amount of work has been done on these genecological aspects in our country (Ramakrishnan, 1960 a, b, c; Pandeya *et. al.*, 1973). Intraspecific variation in *Themeda quadrivalvis* was reported by Pandey and Duckwale (1979) for various traits. These variations were assumed to be due to large ecological amplitude, tolerance of the species, continued hybridization, back crossing and genetical differences.

Turesson (1922a, b) and Anderson and Humbricht (1938) demonstrated that the genetical behaviour of a species can be modified by hybridization and natural selection of accumulated adaptive characters, which can influence hereditary mechanisms, and the variations thus obtained are purely genecological in nature. The process of introgression, if continued, introduces adaptive genes from one species to another, ultimately leading to ecotypic differentiation of taxa.

From the results presented here, it is clear that significant morphological variations exist among the different forms of *D. aegyptium*. It is also borne out by Anderson's scattered diagram, which shows that these polymorphic forms can be distinguished into five complexes, as has already been pointed out. All this indicates the possibility of introgressive hybridization between the forms of one adaptive peak and the other, making possible the formation of new gene combinations that are capable of climbing new adaptive peaks and thus might result in diverse chromosomal races.

The populations of C_1 and C_2 and C_3 have reduced height and total growth vigour, and also suffered high mortality owing to density-dependent competition. This situation can be compared to certain forms of *Agropyron spicatum* that colonize nearby open habitats and are different from those in closed communities (Daubenmire, 1960). Owing to dense plant cover of the grassland community, selection might have eliminated the erect tiller forming growth habit in C_4 and favoured rosette cushion type rhizomatous forms as in C_5. Similar observations were made in *Vernonia patula* with regard to erect and decumbent ecotypes by Jain (1974) and in *Euphorbia hirta* by Ramakrishnan (1960b). Hierarchical development of plants in even aged stands was also shown by Stern (1965) for *Trifolium subterraneum*.

According to Heslop-Harrison (1964), phenotypic interpopulation variation may stem from genetic divergence of selection. The prostrate decumbant type of *Vernonia patula* under constant pressure of grazing and trampling on compact soil surface of open grasslands has adapted

Table 5. Mean and Standard Deviation of certain characters of C_4, C_5 and L_2 of *D. aegyptium* grown in reciprocal sites

Character	C_4			C_5			L_2		
	Home-site	Away-sites		Home-site	Away-sites		Home-site	Away-sites	
	C_4	C_5	L_2	C_5	C_4	L_2	L_2	C_4	C_5
Length of root	14.20±0.60	14.40±0.61	14.30±0.60	18.36±0.70	17.63±0.72	17.90±0.69	9.40±0.42	8.60±0.61	8.80±0.58
Length of shoot	82.30±3.10	78.10±4.10	80.00±3.46	32.30±1.60	30.60±1.20	31.00±1.40	41.32±1.10	38.32±0.92	39.89±0.94
Number of tillers	11.10±0.81	10.40±0.90	10.00±0.83	8.60±0.82	8.10±0.80	8.00±0.80	3.27±0.32	3.10±0.34	3.20±0.33
Number of leaves	83.90±4.20	79.70±5.10	81.20±4.80	139.20±10.60	125.40±9.80	117.70±11.10	28.44±1.42	24.82±1.48	26.10±1.41
Number of inflorescences	10.40±0.96	9.60±0.61	9.80±0.89	11.10±1.40	10.10±1.60	10.60±1.70	4.10±0.52	3.90±0.54	3.70±0.55
Number of spikelets/spike	64.20±1.35	62.80±1.53	62.10±1.31	37.33±1.20	33.73±1.10	32.10±1.20	48.89±1.13	46.98±1.31	44.85±1.38

itself to reduced growth habit (Jameson, 1963). Exact correlation can be drawn from the present populations of C_5 (Fig. 1) with mat-forming structure resulting from biotic influence. If the plants have to adapt themselves to poorer conditions, they tend to be decumbent like C_3 and dwarf like C_2 as was also observed in *Plantago maritima* (Gregor, 1946), *Euphorbia hirta* (Ramakrishnan, 1960b), *E. thymifolia* (Ramakrishnan, 1965a,b) etc. The present results indicate the possible derivation of C_5 from C_4 in response to a change in morphological and physiological characters brought about by grazing, as suggested for herbs by Daubenmire (1974), Jameson (1963), and others. Stapledon (1928) recognized a number of biotic ecotypes in *Dactylis glomerata* in relation to grazing, and so did Ramakrishnan (1960b) in *Euphorbia hirta*. Thus C_5 population can be considered as a biotic ecotype.

The extent of variability observed among compact and lax forms is qualitative, and to some extent quantitative as well, and can be related to the genetic diversity, resulting in a variety of phenotypes, which are adapted to different spatial or temporal phases in the community. According to Sokal (1962), most of the characters that are considered as components of fitness in a population are quantitative characters, and the natural selection should affect them more than any other characters. These characters are expected to respond rapidly and vigorously even to minor changes in environmental conditions. Thus some of the polymorphic forms/ecads in *D. aegyptium* (C_1, C_2, C_3, L_1, L_3) are the products of variation induced by environment. Hence the differentiation of these polymorphic forms may be due to the evolution of phenotypic plasticity that enables the individuals to react during their development in an adaptive fashion to the specific conditions they experience, and the direction followed by a particular evolving population may depend on the genetic archetype of the species (Chinnappa, 1985).

So in the present grassland composed of dominant grasses like *Cymbopogon, Chloris, Echinocloa, Heteropogon* etc., growing densely and absorbing large part of light, the success of the plants depends on their growth to a high level in the canopy while suppressing the neighbours. Thus forms like C_5 and L_2 are resultant of group selection due to heterogeneity of available resources and their neighbouring associates (*Chloris, Cymbopogon* etc.) that might have regulated this species diversity in a frequency-dependent manner.

REFERENCES

Anderson, E. 1957. A semigraphical method for the analysis of complex problems. *Proc. Nat. Acad. Sci. Wash.,* **43**: 923–927.

Anderson, E. and Hubricht, L. 1938. The evidence for introgressive hybridization. *Am. J. Bot.* **25**: 396–402.

Chinnappa, C.C. 1985. Biosystematics of the *Stellaria longipes* complex (Caryophyllaceae). *The J. Cytol. and Genet.,* **20**: 46–58.

Clauson, J., Keck, D.D. and Hesey, W.M. 1948. Experimental studies on the nature of the species. III. Environmental responses of climatic races of *Achillea*. Publ. Carnegie Inst. Washington. 581: 129 pp.

_______. 1940. Experimental studies on the nature of species. I. The effect of varied environments on Western-North American Plants. Publ. Carnegie Inst., Washington, **520**: 1–452.

Daubenmire, R.F. 1960. An experimental study of variation in the *Agropyron spicatum–A. inerma* compex. *Bot. Gaz.,* **122**: 104–108.

_______. 1974. Plants and Environment. A text book of Plant autecology. Wiley Eastern Private Ltd., New Delhi.

Gregor, J.W. 1946. Ecotypic differentiation. *New Phytol.,* **45**: 254–270.

Hanson, H.C. and Churchill, E.D. 1961. The Plant Community, Reinhold Pub. Corporation, New York.

Harper, J.L. 1970. Dynamics of Population. In: Readings in Population and Community Ecology. ed. Hazen, W.E., W.B. Saunders Co., Philadelphia.

Heslop-Harrison, J. 1964. Forty years of genecology. *Advances in Ecological Research,* **2**: 159–247.

Jain, N.K. 1974. Studies in two ecotypes of *Vernonia patula* (Dryand) Merrill. I. Morphology and Physiology. *J. Indian Bot. Soc.,* **53**: 88–96.

Jain, S.K. and Bradshaw, A.D. 1966. Evolutionary divergence among adjacent plant populations. I. The evidence and its theoretical analysis. *Heredity,* **21**: 407–441.

Jameson, D.A. 1963. Responses of individual plants to harvesting. *Bot. Rev.,* **29**: 532–594.

Langlet, O. 1971. Two hundred years of genecology. *Taxon.* **20**: 653–722.

Mary, T.N. and Vijaya Koteswari, M. 2003. The Studies on genetic variability in crowfoot grass (*Dactyloctenium aegyptium* (L.) Beauv.) [Poaceae] from Guntur, Andhra Pradesh. In: T. Pullaiah (ed.) *Biodiversity in India,* **2**: 1–53.

Mayor, E. 1970. Populations, speciation and evolution. The Balknap Press of Harward University, Cambridge, Massachusetts.

Pandey, H.N. and Duckwale, R.N. 1979. Infraspecific variation in *Themeda quadrivalvis* populations. *Geobios.,* **6**: 107–111.

Pandeya, S.C., Sharma, S.C., Pathak, S.J., Jain, H.K., Paliwal, K.C. and Bhanot, V.M. 1973. Autecology and genecology of Anfan grass *Cenchrus ciliaris* complex. Project report No. A7-CR-310 (US PL 480) Dept. of Biosciences, Saurashtra Univ., Rajkot.

Ramakrishnan, P.S. 1960a. Distribution of *Euphorbia thymifolia* Linn. In relation to soil calcium. *Man. Indian Bot. Soc.,* **3**: 52–57.

_______. 1960b. Studies in the autecology of *Euphorbia hirta* Linn., *J. Indian Bot. Soc.,* **39**: 455–473.

_______. 1960c. Ecology of *Echinocloa colonum* Linn., *Proc. Indian Acad. Sci.,* **52**: 73–90.

Sokal, R.R. 1962. Some stages in the development of the concept of natural selection. *Univ. Kensas Sci. Bull.* **42**: 129–151.

Stapledon, R.G. 1928. Cocksfoot grass (*Dactylis glomerata* L.) ecotypes in relation to biotic factors. *J. Ecol.,* **16**: 71–104.

Stern, W.R. 1965. The effect of density on the performance of individual plants in subterranean clover swards.. *Aust. J. Agric. Res.,* **16**: 541–555.

Turesson, G. 1922a. The species and variety as ecological units. *Hereditas,* **3**: 100–113.

_______. 1922b. The genotypical response of the plant species to the habitat. Hereditas **3**: 211–350.

CHAPTER 3

ENDEMIC FLOWERING PLANT SPECIES OF MAHARASHTRA AND THEIR POSSIBLE UTILIZATION

S.P. Gaikwad and S.R. Yadav***

We seem to be entering into a phase of mass extinction loosing at least one flowering plant species per day from tropical forest alone. Any species once lost is lost forever. Therefore, it is very necessary to conserve biological diversity as much as possible. There is an urgent need to survey the plants of atleast important regions, determine their present status and take appropriate steps in conservation of these plants. Efforts to know the potential values of endemic species, their cultivation, utilization, domesticatoin and introduction into horticulture trade seems to be a practical way of conservation. The extensive and intensive field trips in search of rare and endemic flowering plants of Maharashtra for last decade have revealed that there are number of endangered endemic angiosperms which need immediate attention for their conservation. Anthropogenic pressures leading to modification of habitats, replacement of indigenous species by exotic weeds and heavy grazing have become a major threat to endemic plants of restricted distribution. Based on field visits, the present paper enumerates endemic flowering plant species of Peninsular India that occur in Maharashtra and their present status. Number of these neglected endemic plants could be put to use as ornamentals, medicines and some of them are of commercial value. Propagation, domestication and cultivation of these endemic plants

* Department of Botany, Shivaji University, Kolhapur-416 004
** Department of Botany, Delhi University, Delhi-110 007

have two fold benefits of sustainable utilization and conservation of biodiversity. It is impossible to protect every habitat of these endemic plants. Therefore, the only meaningful alternative for their conservation is to propagate, domesticate and introduce into horticultural trade.

The word 'Endemic' is described to any taxonomic unit or taxon, which occurs in a restricted area usually isolated by geographical or temporal barriers (Ahmedullah and Nayar, 1986). Of the estimated 17000 species of flowering plants in India, 5725 (33.5%) are found to be endemic to India (Nayar, 1996). The three megacentres of endemism in India include Eastern Himalayas, Western Himalayas and Western Ghats. Western Ghats, a major centre of endemism provides shelter for over 4000 species of flowering plants of which about 1500 species are endemic (Nayar, 1986). The State of Maharashtra possesses some 3869 species including intra specific taxa of flowering plants (Singh and Karthikeyan, 2000). Over 678 species including intraspecific taxa endemic to Peninsular India occur in Maharashtra of which about 139 species and 13 intra specific taxa are known only from the State. It is well understood that most of the endemic species have narrow range of distribution and fall under some IUCN category of threatened plants.

Previous Work

The concept of endemism is quite old. A.P. De Candolle in 1855 used term endemic to define as an area of a taxonomic unit, especially species and Engler in 1882 recognized three types of endemism. Several researchers have worked on the concept of endemism using different criterias (Drude, 1890; Briquet, 1905; Diels, 1909; Herzog, 1926; Chevalier and Guenot, 1925; Wulff, 1950; Bramwell, 1972 and Richardson, 1978).

In India serious studies on endemic plants started only since 1969, however, before that some authors used the term endemic or apperantly endemic to give distribution of plant species. Cooke (1901–1908) in his flora of Presidency of Bombay used the term apparently endemic while giving distribution data of some plants. Chatterjee's (1940) pioneering work highlights the position regarding the endemic plants in India. Srinivasan (1959) listed some of the plants need protection. He enumerated endemic, apparently endemic, rare, monotypic genera and economic important plants of the country. Important publications on rare, endangered and endemic plant species of India include that of Subramanyam and Sreemadhavan (1969), Blasco (1971), Rao (1972), Subramanyam and Nayar (1974), Nayar (1977), Raghavan and Singh (1983), Vajravelu and Daniel (1983), Jain and Rao (1983), Raghavan and Singh (1984), Singh and Raghavan (1986), Mistry and Almeida (1989), Nayar and Sastry (1987–1990) and Yadav (1997a and 1997b).

Ahmedullah and Nayar (1886) have brought out first authentic book on endemic plants of Peninsular India entitled 'Endemic Plants of Indian Region, Peninsular India' in which they have enumerated 57 genera and 1940 species including intraspecific taxa endemic to the region with brief note on their distribution. Recently Nayar (1996) published book entitled 'Hot Spots of Endemic Plants of India, Nepal, and Bhutan, in which he hightlighted many aspects related to the threatened and endemic plants. In 2001 Mishra and Singh published a book 'Endemic and Threatened Flowering Plants of Maharashtra' in which they have provided up to date consolidated account of 215 endemic and threatened plants. Of these 215 taxa, 130 species and 12 infraspecific taxa are strictly endemic to Maharashtra and remaining 73 taxa are endemic to Peninsular India.

Plan of Work Done and Methods Adopted

In search of endemic flowering plant species field trips of 1–3 days duration were undertaken to the localities of their occurrence. Careful observations were made on their habitat, distribution, present status and month of flowering and fruiting. Similarly, observations were made on possible uses of these plants as ornamental, medicinal, commercial plants. Local enquiries were made to know their local uses. Plant specimens collected were processed for herbarium by using usual technique and deposited in Herbarium of Shivaji University, Kolhapur. The endemic plant species were photographed in their natural habitat to indicate their ornamental value.

Discussions

Of the total 1788 species and 144 infraspecific taxa of flowering plants endemic to Peninsular India, about 678 species including intraspecific taxa occur in Maharashtra.

Table 1: An account of taxa endemic to Peninsular India that occur in Maharashtra

Name of the Taxa	Dicots	Monocots	Total
Families	62	15	77
Genera	208	96	304
Species including intraspecific taxa	409	269	678

Table 2: Families with high number of endemic species in Maharashtra

DICOTS		MONOCOTS	
Name of the Family	Number of species including intraspecific taxa	Name of the Family	Number of species including intraspecific taxa
Fabaceae	68	Poaceae	110
Acanthaceae	50	Orchidaceae	39
Asclepiadaceae	30	Eriocaulaceae	29
Asteraceae	28	Liliaceae	20
Rubiaceae	21	Cyperaceae	16

Table 3: Genera with high number of endemic species in Maharashtra

DICOTS		MONOCOTS	
Genus	Number of species including intraspecific taxa	Genus	Number of species including intraspecific taxa
Ceropegia	17	*Eriocaulon*	29
Crotolaria	15	*Habenaria*	14
Smithia	10	*Ischaemum*	14
Alysicarpus	9	*Dichanthium*	12
Impatiens	8	*Glyphocloa*	11

Table 4: Monotypic endemic genera in Maharashtra

Sr. No.	Botanical Name	Family
1.	*Dicoelospermum ritchiei* (Chakr.) Jeffrey	Cucurbitaceae
2.	*Frerea indica* Dalz.	Asclepiadaceae
3.	*Seshageria sahyadrica* Ansari & Hem.	Asclepiadaceae
4.	*Danthonidium gammiei* (Bhide) Hubb.	Poaceae
5.	*Hubbardia heptanueron* Bor	Poaceae
6.	*Indopoa paupercula* (Stapf) Bor	Poaceae
7.	*Nogra dalzellii* (Baker) Merr.	Fabaceae
8.	*Pogonachne racemosa* Bor	Poaceae
9.	*Pseudodichanthium serrafalcoides* (Cooke & Stapf) Bor	Poaceae
10.	*Tribolachne cookie* (Stapf) Schenck ex Henr.	Poaceae
11.	*Triplopogon ramosissimus* (Hack.) Bor	Poaceae

Table 5. Genera endemic to Peninsular India that occur in Maharashtra

Sr. No.	Name of the species	Family
1.	*Bhidea* Stapf	Poaceae
2.	*Bonnayodes* Blatt. & Hallb.	Scrophulariaceae
3.	*Carvia* Bremek.	Acanthaceae
4.	*Danthonidium* Hubb.	Poaceae
5.	*Dicoelospermum* Cl.	Cucurbitaceae
6.	*Erinocarpus* Nimmo ex J. Grah.	Tiliaceae
7.	*Frerea* Dalz.	Asclepiadaceae
8.	*Gantelbua* Bremek.	Acanthaceae
9.	*Glyphochloa* Clayton	Poaceae
10.	*Helicanthes* Danser	Loranthaceae
11.	*Hubbardia* Bor	Poaceae
12.	*Indopoa* Bor	Poaceae
13.	*Lamprachaenium* Benth.	Asteraceae
14.	*Moullava* (Rheede) Adanson	Caesalpinaceae
15.	*Nanothamnus* Thoms.	Asteraceae
16.	*Nilgirianthus* Bremek	Acanthaceae
17.	*Pleocualis* Nees.	Acanthaceae
18.	*Pogonachne* Bor	Poaceae
19.	*Polyzygus* Dalz.	Apiaceae
20.	*Pseudodichanthium* Bor	Poaceae
21.	*Seshagiria* Ans. & Hem.	Asclepiadaceae
22.	*Supushpa* Suryan.	Acanthaceae
23.	*Trilobachne* Schnek ex Henr.	Poaceae
24.	*Triplopogan* Bor	Poaceae

Potential Values of some Endemic Flowering Plant Specis in Maharashtra

It is said that there is no plant species which is useless, however, probable potential values of endemic plants of Maharashtra based on field observations and information collected from published literature and local people is given below:

1. *Plants of Commercial Value*

The seeds and corms of *Camptorrhiza indica, Iphigenia indica, I. magnifica* and *I. stellata* contain colchicine and could be a commercial tropical source of colchicine. Of them, *Iphigenia stellata* is one of the most promising sources of colchicine. There is need to find out cultivation practices for commercial exploitation of colchicine from these plants. The bulbs of *Drimia* species contain alkaloids used in heart troubles. In Maharashtra, there are 5 species of *Drimia* of which *D. coromandelliana* (tetraploid of *D. indica* Jessop) from Kolhapur region contains highest glycosides and is promising source of glycosides.

Abutilon ranadei Woodr. and Stapf.

Clematis heynei Rau

Delphinium malabaricum (Huth) Munz var. *malabaricum*

Dechaschistia trilobata Wight

PLATE-I

Impatiens dalzellii Hook. f. & Thoms.

Impatiens lawii Hook. f.

Impatiens pulcherrima Dalz.

Impatiens scabrisculata Heyne ex Roxb.

PLATE-II

Alysicarpus belgaumensis Wight

Alysicarpus pubescens Laws. var. *vasavadae* (Hemadri) Sanj.

Spenostylis bracteatus (Baker) Gillet

Vigna khandalensis (Sant.) Raghavan & Wadhwa

PLATE-III

Flemingia nilgheriensis (Baker) Wight ex Cooke

Nogra dalzellii (Baker) Merr.

Indigofera dalzelli T. Cooke

Smithia agharkarii Hemadri

Smithia setulosa Dalz.

PLATE-IV

Bidaria khandalense (Sant.) Jagtap & Singh

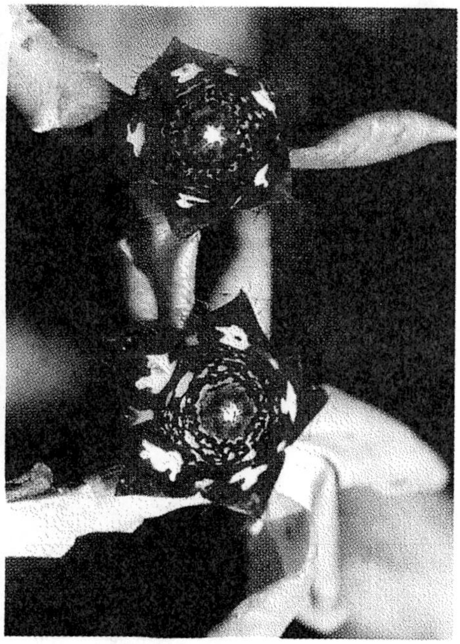

Frerea indica Dalz.

Ceropegia noorjahaniae Ansari

Ceropegia panchganiensis Blatt. & McC

PLATE-V

Barleria cuspidata Heyne ex Nees

Barleria gibsoni Dalz.

Pleocaulis ritchiei (Cl.) Bremek.

Dicliptera nasikensis Lakmi. & Sharma

Barleria grandiflora Dalz.

PLATE-VI

Crinum brachynema Herb.

Drimia razii Ansari

Camptorrhiza indica Yadav, Singh & Mathew

Trithuria konkanensis Yadav & Janarthanam

Dipcadi concanense (Dalz.) Baker

PLATE-VII

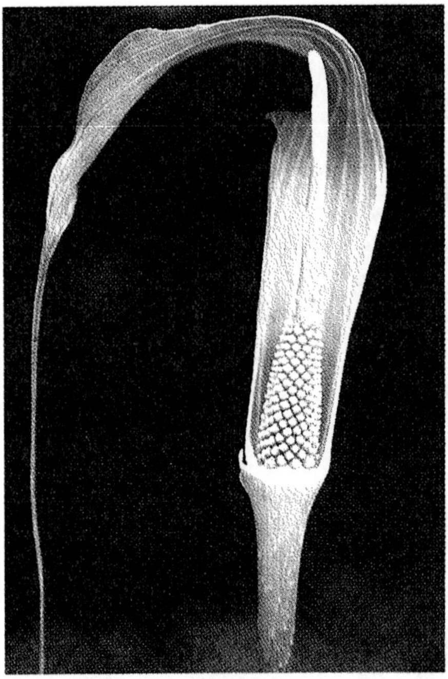

Arisaema caudatum Engl.

Arisaema murrayii (Grah.) Hook. f.

Arisaema sivadasanii Yadav, Patil & Janarthanam

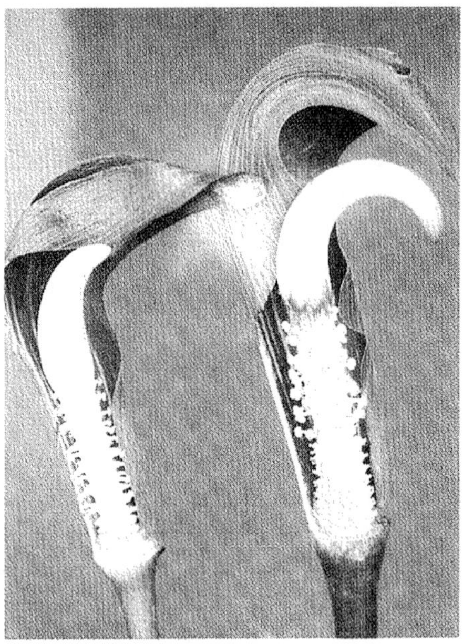

Arisaema sahyadricum Yadav, Patil & Bachulkar

PLATE-VIII

Cultivation and improvement of strains through genetic manipulation is worth trying in commercial productions of glycosides.

2. *Plants of Medicinal Value*

Plants like *Ancistrocladus heyneanus, Delphinium malabaricum, Hydnocarpus pentandra, Moullava spicata, Holigarna arnottiana, H. grahamii, Salacia brunoniana* and species of *Heracleum* and *Zingiber* etc. are of medicinal value and so far the plant materials of medicinal plants are mainly exploited from wild sources. There is need to cultivate at least important rare medicinal plants.

3. *Plants of Food Value*

Cucumis setosus, Flemingia tuberosa, Flemingia nilgheriensis, Merremia rhyncorhiza and some species of *Amorphophallus, Brachystelma, Ceropegia, Chlorophytum* etc. have tuberous edible rootstock of some food value. Some of them could be introduced as tuberous crops.

4. *Plants of Ornamental Value*

There is a long list of endemic plants of Maharashtra, which could be used as ornamental plants for various purposes in gardens. Plant species that can be introduced into gardens as evergreen trees include *Arenga wightii, Calophyllum apetalum, Dimorphocalyx lawianus, Euonymus indicus, Garcinia talbotii, Gordonia obtusa, Knema attenuata, Maba nigrescens, Memecylon talbotianum, Meiogyne pannosa, Myristica malabarica, Pittosporum dasycaulon, Phoenix robusta, Sageraea laurifolia, Syzygium laetum.* Some of the endemic shrubs, which have large beautiful flowers of ornamental value *viz. Abutilon ranadei, Argyeia cuneata, Barleria cuspidata, B. gibsoni, B. gibsonoides, B. grandiflora, B. involucrata* var. *elata, Beaumontia jerdoniana, Calacanthus grandiflorus, Carvia callosa, Decaschistia trilobata, Ochna obtusata, Pleocaulis ritchiei* and *Supushpa scrobiculata* which deserve place in gardens. Many of the endemic herbaceous plants have flowers of ornamental value and could be introduced in gardens as seasonals. Important herbaceous plants known for elegant flowers include species of *Aerides, Bigonia, Crinum, Curcuma, Cyanotis, Delphinium, Dendrobium, Habenaria, Impatiens* and *Smithia, Paracaryum malabaricum, Pinda concanensis* and many others.

Endemic aquatic plants such as *Aponogeton bruggenii, A. satarensis, Cryptocoryne cognata, E. setaceum, Nymphoides aurantiacea, Rotala ritchiei* etc. have ornamental value and could be used as ornamentals for aquaria.

5. Plants of Botanical Interest

Number of rare plants of Maharashtra are of botanical interest. Although uses of most of the rare plant species such as *Bhidea bursiana, Danthonidium gammiei, Frerea indica, Indopoa paupercula, Pseudodichanthium serrafalcoides, Seshagiria sahyadrica, Trithuria konkanensis, Triplopogon ramosissimus, Trilobachne cookei, Wiesneria triandra* etc. are presently not well understood but they have great importance in understanding biogeography and evolution of plants. In addition to potential values of rare plants of Maharashtra, some of the wild plants form important genetic pool to improve our timber, fibers, food and fruit plants through hybridization and selection of desired genes for disease resistance, food value and yield. Wild germplasm of *Asparagus, Crinum, Cucumis, Curcuma, Garcinia, Vigna* and *Zinziber,* members of Poaceae, Fabaceae and Vitaceae etc. is of great significance in plant breeding programs and human welfare.

From field observations and literature survey, it is evident that most of the indigenous rare plants have medicinal, commercial, ornamental or at least botanical importance. In past, no efforts have been made to find out the potential values of rare plants and they remained neglected due to their rarity. There is urgent need to take some definite steps for their conservation and utilization in sustainable development of the country.

Acknowledgement

Authors are thankful to Registrar, Department of Botany, Shivaji University Kolhapur, Maharashtra, India for providing research facilities; to Pragat Prathistan Bombay for partial financial assistance and first author is thankful to Department of Science and Technology, Govt. of India for financial assistance under SERC Fast Track Scheme for Young Scientist 2000.

Taxonomic Account

Enumeration of endemic flowering plant species of Maharashtra with their month of flowering and fruiting and present status in the State. *Taxa so far known from only Maharashtra.

Plant Name	*Month of Flowering and fruiting*	*IUCN Red List Category*
DICOTYLEDONS		
RANUNCULACEAE		
Clematis heynei Rau	Sept. – Nov.	Low Risk
**Delphinium malabaricum* (Huth.) Munz. var. *malabaricum*	Aug. – Oct.	Vulnerable
**D. malabaricum* (Huth.) Munz. var. *ghaticum* Billore	Sept. – Nov.	Extinct (?)
Thalictrum dalzellii Hook.	Aug. – Sept.	Endangered
**T. obovatum* Blatt.	Aug. – Sept.	Extinct (?)
ANNONACEAE		
Meiogyne pannosa (Dalz.) Sinclair	Oct. – Apr.	Low Risk
Polyalthia fragrans (Dalz.) Bedd.	Oct. - May	Vulnerable
Sageraea laurifolia (Grah.) Blatt.	Oct. - Apr.	Low Risk
CAPPARIDACEAE		
Capparis grandiflora Wall. ex Hook. f. & Thoms.	Mar. – Apr.	Low Risk
C. rheedei DC.	Mar. - Apr	Data Deficient
Cleome felina L.	Aug. –Sept.	Not Evaluated
C. simplicifolia Hook. f. & Thoms.	Mar. – Dec.	Low Risk
FLACOURTIACEAE		
Casearia rubescens Dalz.	Feb. - May	Not Evaluated
Flacourtia latifolia (Hook. f. & Thoms.) Cooke	Mar. – Jul.	Low Risk
Flacourtia montana Grah.	Nov. – Apr.	Low Risk
Hydnocarpus pentandra (Buch.-Ham) Oken.	Mar. – Aug.	Low Risk
PITTOSPORACEAE		
Pittosporum dasycaulon Miq.	Nov. – Jan.	Low Risk
CARYOPHYLLACEAE		
Polycarpaea corymbosa (L.) Lam. var. *aurea* (Wight & Arn.) Wight	Aug. – Jan.	Data Deficient
CLUSIACEAE		
Calophyllum apetalum Willd.	Dec. – Mar.	Vulnerable
Garcinia gummi-gutta (L.) Robson var. *gummi-gutta*	Jan. – Mar.	Not Evaluated
G. gummi-gutta var. *conicarpa* (Wight) Singh	Mar. – Jan.	Not Evaluated
G. indica (Du Petit-Thou.) Choisy	Nov. – May	Low Risk
G. talbotii Raiz.	Dec. – May	Low Risk

(contd....)

Plant Name	*Month of Flowering and fruiting*	*IUCN Red List Category*
THEACEAE		
Gordonia obtusa Wall. ex Wight & Arn.	May – Jan.	Not Evaluated
DIPTEROCARPACEAE		
Hopea ponga (Dennst.) Mabb.	Mar. – Sept.	Not Evaluated
ANCISTROCLADACEAE		
Ancistrocladus heyneanus Wall. ex Grah.	Jan. – Jun.	Low Risk
MALVACEAE		
Abelmoschus manihot (L.) Medik. ssp. *tetraphyllus* (Roxb. ex Horn.) Borss. var. *megaspermus* Hemadri	Oct. – Jan.	Not Evaluated
**Abutilon ranadei* Woodr. & Stapf	Nov. – Mar.	Endangered
Decaschistia crotonifolia Wall. ex Wight & Arn.	Mar. – Nov.	Data Deficient
D. trilobata Wight	Oct. – Jan.	Vulnerable
Hibiscus talbotii (Rasshit) Paul & Nayar	Nov. – Jun.	Low Risk
STERCULIACEAE		
Eriolaena quiquelocularis (Wight & Arn.) Wight	May – Jul.	Low Risk
E. stocksii Hook. f. & Thoms.	Jul. – Oct.	Data Deficient
TILIACEAE		
Erinocarpus nimmonii Grah.	Sept. – Dec.	Low Risk
Grewia heterotricha Mast.	Apr. – Jan.	Low Risk
G. umbellifera Bedd.	Oct. – Apr.	Low Risk
ELAEOCARPACEAE		
Elaeocarpus munroii (Wight) Mast.	Sept. – Nov.	Data Deficient
MALPIGHIACEAE		
Aspidopteris canarensis Dalz.	Feb. – Mar.	Not Evaluated
A. cordata (Heyne ex Wall.) Juss.	Aug. – Nov.	Low Risk
ZYGOPHYLLACEAE		
Tribulus rajasthanensis Bhandari & Sharma	Aug. – Dec.	Low Risk
T. subramanyamii Singh, Giri & Singh	Jul. – Aug.	Data deficient
BALSAMINACEAE		
Impatiens dalzellii Hook.f. & Thoms.	Aug. – Sept.	Low Risk
I. inconspica Benth. ex Wight	Aug. – Sept.	Low Risk
I. latifolia L.	Aug. – Sept.	Low Risk

(contd....)

Plant Name	*Month of Flowering and fruiting*	*IUCN Red List Category*
I. lawii Hook. f. & Thoms.	Aug. – Sept.	Low Risk
I. minor (DC.) Bennet	Aug. – Sept.	Low Risk
I. pulcherrima Dalz.	Aug. – Sept.	Low Risk
I. scabrisculata Heyne ex Roxb.	Aug. – Oct.	Low Risk
I. tomentosa Heyne ex Wight & Arn.	Aug. – Sept.	Low Risk
RUTACEAE		
Atalantia wightii Tanaka	Nov. – Feb.	Data Deficient
Luvunga eleutherandra Dalz.	Jan. – Jun.	Low Risk
OCHNACEAE		
Ochna obtusata DC var. *gamblei* (King ex Brandis) Kanis	Apr. – Jun.	Not Evaluated
BURSERACEAE		
Boswellia serrata Roxb. ex Colebr.	Feb. – Jun.	Low Risk
MELIACEAE		
Naregamia alata Wight & Arn.	Aug. – Dec.	Low Risk
Reinwardtiodendron anamalaiense (Bedd.) Mabb.	Mar. – Nov.	Data Deficient
Toona microcarpa (DC.) Harms var. *pilipetala* (DC.) Bahadur	Apr. – Jul.	Data Deficient
AQUIFOLIACEAE		
Ilex malabarica Bedd.	Feb. – Apr.	Data Deficient
CELASTRACEAE		
Euonymus indicus Heyne ex Wall.	Jan. – Oct.	Not Evaluated
Maytenus puberula (Laws.) Loes	Jul. – Sept.	Not Evaluated
**Salacia brunoniana* Wight & Arn.	Jan. – Feb.	Critically endangered
RHAMNACEAE		
Rhamnus purandharensis Wight & Arn.	Feb. – May	Not Evaluated
Smythea bombaiensis	Dec. – Jun.	Low Risk
Ventilago gamblei Suesseng.	Feb. – Jul.	Not Evaluated
**V. madraspatana* Gaertn. var. *fructifida* Sant.	Dec. – Jan.	Critically Endangered
Ziziphus horrida Roth	Dec. – Jan.	Data Deficient
**Z. rugosa* Lam. var. *glabra* Bhandari & Bhansali	Jan. – Feb.	Data deficient
Z. rugosa Lam. var. *glabrescens* Bhandari & Bhansali	Jan. – Feb.	Data Deficient
Z. walliamani Bhandari & Bhansali	Nov. – Mar.	Data Deficient
VITACEAE		
Ampelocissus araneosa (Dalz. & Gibs.) Planch.	Sept. – Oct.	Data Deficient

(contd....)

Plant Name	*Month of Flowering and fruiting*	*IUCN Red List Category*
Cissus latifolia Lam.	May – Sept.	Not Evaluated
C. woodrowii (Stapf ex Cooke) Sant.	Jun. – Oct.	Low Risk
SAPINDACEAE		
Allophylus cominia (L.) Sw.	Apr. – Sept.	Not Evaluated
ANACARDIACEAE		
Holigarna arnottiana Hook. f.	Jan. – Jun.	Low Risk
H. grahamii (Wight) Kurz.	Dec. – Jun.	Low Risk
FABACEAE		
Alysicarpus belgaumensis Wight	Sept. – Dec.	Low Risk
**A. narimanii* Almeida & Almeida	Sept. – Oct.	Data deficient
A. pubescens Laws. var. *vasavadae* (Hemadri) Sanj.	Sept – Dec.	Low Risk
A. racemosus Benth.	Sept. – Nov.	Low Risk
A. roxburghianus Thoth. & Pram.	Sept. – Nov.	Low Risk
**A. salim-ali* Almeida & Almeida	September.	Data Deficient
A. tetragonolobus Edgw. var. *pushanensis* Almeida & Almeida	November	Data Deficient
Cajanus lineatus (Wight & Arn.) van der Masen	Oct. – Apr.	Low Risk
C. sericeus (Benth. ex Baker) van der maesen	Oct. – Apr.	Low Risk
Clitoria biflora Dalz.	Oct. – Apr.	Low Risk
C. ternatea L. var. *pilosula* Wall ex Baker	Jul. – Oct.	Data Deficient
**Crotalaria decasperma* Naik	Sept. – Nov.	Vulnerable
C. epunctata Dalz.	Dec. – Feb	Data Deficient
C. filipes Benth. var. *filipes*	Aug. – Feb.	Low Risk
C. filipes Benth. var. *trichophora* (Benth. ex Baker) Cooke	Aug. – Feb.	Low Risk
C. leptostachya Benth.	Oct. – Nov.	Low Risk
C. lunulata Heyne ex Wight & Arn.	Not Known	Data Deficient
C. lutescens Dalz.	Aug. – Dec.	Low Risk
C. madurensis Wight & Arn.	Dec. – Feb.	Data Deficient
C. nakiana Zate	Sept. – Nov.	Not Evaluated
C. notonii Wight & Arn.	Sept. – Dec.	Low Risk
C. priestleyoides Benth. ex Baker	Sept. – Dec.	Data deficient
C. pusilla Heyne ex DC.	Aug. – Dec.	Low Risk
C. quinquefolia L.	Nov. – Feb.	Data Deficient
C. tecta Roth	Aug. – Oct.	Data Deficient
C. vestita Baker	Sept. – Jan.	Low Risk
Dalbergia horrida (Dennst.) Mabb. *var. horrida*	Nov. – Jun.	Low Risk
D. malabarica Prain	Feb.- Mar.	Data Deficient
Derris brevipes (Benth.) Baker	May – Sept.	Low Risk

(contd....)

Plant Name	*Month of Flowering and fruiting*	*IUCN Red List Category*
D. heyneana (Wight & Arn.) Benth.	Feb. – Jun.	Low Risk
D. thothathrii Bennet	Dec. – Mar.	Data Deficient
Desmodium benthamii Balakr.	Sept. – Nov.	Data Deficient
D. ritchiei Sanj.	Sept. – dec.	Not Evaluated
Flemingia gracillis (Mukerjee) Ali	Sept. – Oct.	Endangered
F. nilgheriensis (Baker) Wight ex Cooke	Sept. –Nov.	Low Risk
F. precox Cl. ex Prain var. *robusta* (Mukerjee) Sanj.	Aug. – Oct.	Data Deficient
**F. rollae* (Billore & Hemadri) Kumar	Sept. – Oct.	Critically endangered
F. tuberosa Dalz.	Sept. – Nov.	Low Risk
Galactia tenuifolia (Klein ex Willd.) Wight & Arn. var. *latifolia* Baker	Sept. – Dec.	Data Deficient
**G. tenuifolia* (Klein ex Willd.) var. *minor* Baker	Sept. – Dec.	Low Risk
Indigofera angulosa Edgew.	Sept. – Oct.	Not Evaluated
I. dalzelli T. Cooke	Sept. – May	Low Risk
**I. deccanensis* Sanj.	Jul. – Nov.	Vulnerable
I. glandulosa Wendl. var. *sykesii* Griff. ex Baker	Sept. – Nov.	Low Risk
I. prostrata Willd.	Sept. – Jan.	Low Risk
**I. santapaui* Sanj.	Aug. – Nov.	Data Deficient
**I. trifoliata* L. var. *duthiei* (Drumm. ex Naik) Sanj.	Aug. – Nov.	Low Risk
**I. trita* L. var. *purandharensis*. Sanj.	Aug. – Nov.	Data Deficient
I. uniflora Buch-ham.ex Roxb.	Aug. – Nov.	Vulnerable
Millettia orissae Panigr.	May – Mar.	Data Deficient
Nogra dalzellii (Baker) Merr.	Aug. – Sept.	Endangered
Sesbania procumbens (Roxb.) Wight & Arn.	Feb. – Mar.	Not Evaluated
**Smithia agharkarii* Hemadri	Sept. – Oct.	Low Risk
S. bigemina Dalz.	Aug. – Jan.	Low Risk
S. capitata Dalz.	Sept. – Oct.	Data Deficient
S. gracilis Benth.	Aug. – Oct.	Data Deficient
S. hirsuta Dalz.	Aug. – Oct.	Low Risk
S. oligantha Blatt.	January	Data Deficient
S. purpurea Hook.	Sept. – Oct.	Low Risk
S. pycnantha Benth. ex Baker	Sept. – Oct.	Data Deficient
S. salsuginia Hance	Sept. – Oct.	Low Risk
S. setulosa Dalz.	Sept. – Oct	Low Risk
Spatholobus purpureus Benth. ex Baker	Dec. – Feb	Low Risk
**Spenostylis bracteatus* (Baker) Gillet	Sept. – Oct.	Vulnerable
Tephrosia canarensis Drumm.	Aug. – Oct.	Data Deficient
T. coccinia Wall.	Sept. – Nov.	Low Risk

(contd....)

Plant Name	*Month of Flowering and fruiting*	*IUCN Red List Category*
T. collina Sharma	Aug. – Oct.	Low Risk
Vigna khandalensis (Sant.) Raghavan & Wadhwa	Aug. – Oct.	Low Risk
V. trilobata (L.) Verdc. var. *pusilla* Naik & Pokle	Aug. – Oct.	Low Risk
V. vexillata (L.) A. Rich. *var. sepiara* (Dalz.) Babu & Sharma	Sept. – Oct.	Data Deficient
V. vexillata (L.) A. Rich. *var. stocksii* Benth. ex Baker	Sept. – Oct.	Data Deficient
CAESALPINACEAE		
Baauhinia foveolata Dalz.	Oct. – Dec.	Not Evaluated
B. phoenicea Heyne ex Wight & Arn.	Nov. – Dec.	Data deficient
**Cassia kolabensis* Kothari, Nayar & Moorthy	Aug. – Oct.	Vulnerable
Hardwickia binata Roxb.	Aug. – Jan.	Low Risk
Moullava spicata (Dalz.) Nicols.	Nov. – Apr.	Low Risk
MIMOSACEAE		
Dichrostachys cinera var. *indica* Bremen & Brumik	Jun. – Oct.	Low Risk
CRASSULACEAE		
Kalanchoe bhidei Cooke	Oct. – Nov	Low Risk
K. olivacea Dalz.	Oct. – Dec.	Low Risk
COMBRETACEAE		
Terminalia paniculata Roth	Apr. – Aug.	Low Risk
MYRTACEAE		
Syzygium laetum (Buch-Ham.) Gandhi	Jan. – May	Vulnerable
S. stocksii (Duthie) Gamble	Jan. – Feb.	Data Deficient
MELASTOMACEAE		
Memecylon talbotianum Brandis	Mar. – Apr.	Low Risk
Sonerilla rheedei Wight & Arn.	Jul. – Sept.	Not Evaluated
S. scapigera Hook.	Jul. – Aug.	Not Evaluated
LYTRACEAE		
Ammania nagpurensis Matthew & Nayar	Oct. – Nov.	Not Evaluated
Lagerstroemia microcarpa Wight	Mar. – Jun.	Low Risk
**Rotala floribunda* (Wight) Koehne	Oct. – Feb.	Low Risk
**Rotala ritchei* (Cl.) Koehne	Sept. – Dec.	Critically Endangered (?)

(contd....)

Plant Name	*Month of Flowering and fruiting*	*IUCN Red List Category*
CUCURBITACEAE		
Corallocarpus conocarpus (Rottl.) Cl.	Aug. – Oct.	Low Risk
Cucumella ritchiei (Chakr.) Jeffery	Mar. – Apr.	Not Evaluated
Cucumis setosus Cogn.	Sept. – Oct.	Low Risk
**Dicoelospermum ritchiei* Cl.	Aug. – Oct.	Low Risk
BEGONIACEAE		
Begonia concanensis DC.	Aug. – Nov.	Vulnerable
B. crenata Dryand.	Aug. – Nov.	Low Risk
**B. phrixophylla* Blatt. & McC.	Aug. – Oct.	Data Deficient
B.. trichocarpa Dalz.	Aug. – Oct.	Low Risk
APIACEAE		
Heracleum aquilegifolium Cl.	Sept. – Oct.	Not Evaluated
**H. dalgadianum* Almeida	Sept. – Oct.	Endangered
H. grande (Dalz. & Gibs.) Mukh.	Jul. – Aug.	Low Risk
H. sprengelianum Wight & Arn.	Aug. – Sept.	Not Evaluated
Peucedanum dhana Cl. *var. dalzellii* Cl.	Aug. – Sept.	Not Evaluated
Pimpinella adscendens Dalz.	Jan. – Feb.	Low Risk
**P. rollae* Billore & Hemadri	Jul. – Nov.	Critically Endangered
**P. tomentosa* (Dalz. & Gibs.) Cl.	Aug. – Sept.	Low Risk
P. wallichiana (Miq ex Hohen.) Gandhi	Sept. – Oct.	Low Risk
**Pinda concanense* (Dalz.) Mukh. & Constance	Sept. – Oct.	Low Risk
Polyzygus tuberosus Dalz.	Aug. – Oct.	Vulnerable
Seseli diffusum (Roxb. ex Sm.) Sant.	Apr. – May	Not Evaluated
Trachyspermum roxburghianum (DC.) Craib.	Aug. – Oct.	Not Evaluated
RUBIACEAE		
Discospermum apiocarpa Dalz.	Jun. – Jan.	Not Evaluated
D. sphaerocarpum Dalz.	Oct. – Dec.	Not Evaluated
Hedyotis stocksii (Hook. f. & Thoms.) Rao & Hamadri	Sept. – Nov.	Not Evaluated
Hymenodictyon obovatum Wall.	Aug. – Mar.	Low Risk
Ixora brachiata Roxb.	Nov. – Feb.	Low Risk
I. elongata Heyne ex G. Don	Feb. – May	Low Risk
I. polyantha Wight	Jan. – Jun.	Low Risk
Lasianthus acuminatus Wight	Jan. – Apr.	Vulnerable
L. sessilis Talb.	Jul. – Sept.	Not Evaluated
L. venulosus (Wight & Arn.)	Feb. – Apr.	Data Deficient
Mussaenda belilla Buch. – Ham.	Apr. – Nov.	Low Risk
Neanotis carnosa (Dalz.) Lewis	Aug. – Sept.	Not Evaluated
N. decipiens (Hook. f.) Lewis	Not Known	Data Deficient
N. foetida (Dalz.) Lewis	Aug. – Oct.	Low Risk

(contd....)

Plant Name	*Month of Flowering and fruiting*	*IUCN Red List Category*
N. lancifolia (Dalz.) Lewis	Not Known	Data Deficient
N. montholonii (Hook.f.) Lewis	Sept. – Oct.	Low Risk
**N. sahyadrica* Billore & Mudaliar	Oct. – Nov.	Critically Endangered
Pavetta siphonantha Dalz.	Apr. – Nov.	Low Risk
Psychotria dalzellii Hook. f.	Jun. – Mar.	Low Risk
P. flavida Talb.	May – Feb.	Not Evaluated
Saprosma glomeratum (Gardn.) Bedd.	Feb. – Oct.	Not Evaluated
ASTERACEAE		
Adenoon indicum Dalz.	Aug. – Dec.	Low Risk
Anaphalis lawii (Hook. f.) Gamble	Sept. – Mar.	Data Deficient
Blepharispermum subsessile DC.	Aug. – Oct.	Data Deficient
Blumea atropurpurea Haines	Dec. – Jan.	Low Risk
B. belangeriana DC.	Feb. – May	Low Risk
B. eriantha DC.	Dec. – May	Low Risk
B. malcolmii (Cl.) Hook. f.	Oct. – May	Low Risk
**B. venkatramanii* Rao & Hemadri	Dec. – Feb.	Data Deficient
Cyathocline lutea Law. ex Wight	Oct. – Jan.	Vulnerable
**C. purpurea* (Buch.-Ham. ex D. Don) O. Ktze. var. *alba* Sant.	Nov. – May	Not Evaluated
**C. purpurea* (Buch.-Ham. ex D. Don) O. Ktze. var. *bicolor* Sant.	Nov. – Mar.	Not Evaluated
Emilia zeylanica Cl. var. *paludosa* Gamb.	Nov. – Dec.	Data Deficient
Gynura nitida DC.	Sept. – Jan.	Low Risk
G. travancorica Sm.	Feb. – May	Data Deficient
Lamhrachaenium microcephalum (Dalz.) Benth.	Sept. – Jan.	Low Risk
Nenothamus sericeus Thoms.	Feb. – May	Vulnerable
**Phyllocephalum hookeri* (Cl.) Uniyal	Nov. – Dec.	Not Evaluated
P. ritchiei (Hook. f.) Narayana	Sept. – Nov.	Not Evaluated
P. scabridum (DC.) Kirkman	Jun. – Dec.	Low Risk
Senecio belgaumensis (Wight) Cl.	Aug. – May	Low Risk
S. dalzellii Cl.	Sept. – Oct.	Low Risk
S. edgeworthii Hook. f.	Aug. – Nov.	Low Risk
S. gibsonii Hook. f.	Apr. – Nov.	Not Evaluated
S. hewrensis (Dalz.) Hook. f.	Aug. – Oct.	Not Evaluated
Tricholepis amplexicaulis Cl.	Aug. – Nov.	Low Risk
T. glaberrima DC.	Aug. – Nov.	Low Risk
T. radicans DC.	Sept. – Feb.	Low Risk
Vernonia indica Cl.	Sept. – Nov.	Low Risk
CAMPANULACEAE		
Campanula alphonsii Wall. ex DC.	Nov. – Mar.	Vulnerable
C. colorata Wall.	Nov. – Dec.	Not Evaluated

(contd....)

Plant Name	*Month of Flowering and fruiting*	*IUCN Red List Category*
SAPOTACEAE		
Isonandra stocksii Cl.	Not Known	Data Deficient
OLEACEAE		
Jasminum malabaricum Wight	Mar. – Jun.	Low Risk
APOCYNACEAE		
Beamontia jerdoniana Wight	Nov. – Dec.	Vulnerable
Carissa inermis Vahl	Jan. – Feb.	Low Risk
Kamettia caryophyllata (Roxb.) Nicols. & Suresh	Nov. – Jan.	Data Deficient
Tabernaemontona alternifolia (Roxb.) Nicols. & Suresh	Mar. – Apr.	Low Risk
ASCLEPIADACEAE		
Bidaria cuspidata (Thunb.) Huber	Sept. – Oct.	Not Evaluated
B. khandalense (Sant.) Jagtap & Singh	Oct. – Mar.	Vulnerable
**Brachystelma malwanense* Yadav & Singh	Mar. – Apr.	Critically Endangered
**B. naorojii* Tetali	Mar. – Apr.	Critically Endangered
Ceropegia attenuata Hook. f.	Jul. – Aug.	Low Risk
**C. evansii* McC.	Jul. – Oct.	Critically Endangered
C. fantastica Sedgw.	Aug. – Sept.	Critically Endangered
**C. huberi* Ansari	Aug. – Oct.	Critically Endangered
**C. jainii* Ansari & Kulk.	Aug. – Mar.	Low Risk
**C. lawii* Hook.f.	Aug. – Sept.	Vulnerable
**C. maccannii* Ansari	Jul. – Oct.	Vulnerable
**C. mahabalei* Hemadri & Ansari	Aug. – Sept.	Critically Endangered
**C. media* (Huber) Ansari	Jul. – Oct.	Vulnerable
**C. noorjahaniae* Ansari	Jul. – Aug.	Vulnerable
C. oculata Hook. f.	Aug. – Sept.	Low Risk
C. odorata Hook. f.	Aug. – Sept.	Data Deficient
**C. panchganiensis* Blatt. & McC.	Jul. – Aug.	Critically Endangered
**C. rollae* Hemadri	Aug. – Sept.	Critically Endangered
**C. sahyadrica* Ansari & Kulkarni	Aug. – Sept.	Vulnerable
**C. santapaui* Wadhwa & Ansari	Aug. – Sept.	Critically Endangered
**C. vincaefolia* Hook. f.	Aug. – Oct.	Low Risk
**Frerea indica* Dalz.	Sept. – Dec.	Critically Endangered

(contd....)

Plant Name	*Month of Flowering and fruiting*	*IUCN Red List Category*
Heterostemma dalzellii Hook. f.	Sept. – Oct.	Vulnerable
H. deccanense (Talb.) Swarupanadan & Mangaly	Sept. – Oct.	Critically Endangered
H. disciflorum (Hook. f.) Swarup & Mangaly	Sept. – Oct.	Vulnerable
H. urceolatum Dalz.	Jul. – Aug.	Critically Endangered
Hoya retusa Dalz.	Jul. – Aug.	Vulnerable
Sarcostemma intermedium Decne.	December	Data Deficient
Seshagrica sahyadrica Ansari & Hemadri	Jul. – Sept.	Low Risk
**Wattakaka lanceolata* (Cooke) Jagtap & Singh	May – Sept.	Vulnerable
LOGANIACEAE		
Strychnos dalzellii Cl.	Dec. – Mar.	Low Risk
GENTIANACEAE		
Canscora concanensis Cl.	Sept. – Dec.	Vulnerable
C. decurrens Dalz.	Sept. – Feb.	Low Risk
**C. diffusa* (Vahl.) R. Br. var. *tetraptera* Naik & Pokle	Oct. – Jan.	Endangered
**C. khandalensis* Sant.	Sept. – Oct.	Data Deficient
C. pauciflora Dalz.	Sept. – Dec.	Low Risk
C. perfoliata Lamk.	Sept. – Jun.	Low Risk
Exacum lawii Cl.	Jul. – Oct.	Low Risk
E. pumilum Griseb.	Aug. – Oct.	Low Risk
Swertia corymbosa Wight var. *elatior* Griseb.	Not Known	Data Deficient
S. densifolia (Griseb.) Kashyapa	Dec. – Jan.	Not Evaluated
S. lawii Burkill	Sept. – Nov.	Low Risk
S. minor (Griseb.) Knobl.	Jul. – Oct.	Low Risk
MENYANTHACEAE		
Nymphoides aurantiacea (Dalz.) O.Kuntze	Sept. – Dec.	Critically Endangered
BORAGINACEAE		
Cordia domestica Forst.	Feb. – Jun.	Low Risk
Heliotropium cornutum Johnst.	Feb. – Jun.	Data Deficient
Paracaryopsis coelesina (Lindl.) Mill.	Sept. – Nov.	Low Risk
P. malabarica (Cl.) Mill.	Sept. – Oct.	Low Risk
CONVOLVULACEAE		
**Argyreia boseana* Sant. & Patel	Aug. – Oct.	Vulnerable
A. cuneata (Willd.) Ker.-Gawl.	Sept. – Dec.	Low Risk
A. lawii Cl.	Aug. – Nov.	Data Deficient
A. pilosa Wight & Arn.	Sept. – Oct.	Data Deficient

(contd....)

Plant Name	*Month of Flowering and fruiting*	*IUCN Red List Category*
A. sericea Dalz.	Aug. – Nov.	Low Risk
Ipomoea clarkei Hook. f.	Sept. – Oct.	Data Deficient
**I. salsettensis* Sant.	Sept. – Oct.	Data Deficient
Merremia rhynchorhiza (Dalz.) Hallier	Sept. – Oct.	Vulnerable
**Operculina tansaensis* Sant. & Patel	Not Known	Data Deficient
SCROPHULARIACEAE		
**Bonnayodes limnophiloides* Blatt. & Hallb.	Not Known	Data Deficient
L. polystachyoides Blatt.	Dec. – Mar.	Endangered
Lindernia estaminodiosa (Blatt. ex Hallb.) Mukherjee	Sept. – Dec.	Not Evaluated
**L. quinqueloba* (Blatt. & Hallb.) Mukh.	November	Data Deficient
Ramphicarpa longiflora (Arn.) Benth.	Aug. – Oct.	Low Risk
Torenia bicolor Dalz.	Jul. – Nov.	Endangered
T. indica Saldhana	Aug. – Oct.	Low Risk
T. lindernoides Saldanha	May – Aug.	Not Evaluated
LENTIBULARIACEAE		
Utricularia albocaerulea Dalz.	Aug. – Sept.	Low Risk
**U. janarthanamii* Yadav, Sardesai & Gaikwad	Aug. – Sept.	Not Evaluated
U. malabarica Janarthanam & Henry	Aug. – Sept.	Low Risk
U. minutissima Vahl	Aug. – Sept.	Low Risk
**U. naikii* Yadav, Sardesai & Gaikwad	Aug. – Sept.	Vulnerable
U. praeterita Taylor	Aug. – Sept.	Low Risk
U. purpurascens Grah.	Aug. – Sept.	Low Risk
GESNERIACEAE		
Aeschynanthus perrottetii DC.	Aug. - Mar.	Vulnerable
ACANTHACEAE		
Asystasia dalzelliana Sant.	Nov. – Mar.	Low Risk
A. mysurensis (Roth) T. And.	Aug. – Dec.	Low Risk
Barleria courtillica Nees	Feb. – May	Data Deficient
B. cuspidata Heyne ex Nees	Sept. – Dec.	Low Risk
B. gibsoni Dalz.	Sept. – Mar.	Low Risk
**B. gibsonioides* Blatt.	Sept. – Dec.	Data Deficient
B. grandiflora Dalz.	Sept. – Dec.	Vulnerable
B. involucrata Nees var. *elata* (Dalz.) Cl.	Oct. – Mar.	Low Risk
**B. prattensis* Sant.	Sept. – Feb.	Data Deficient
B. sepalosa Cl.	Oct. – Feb.	Not Evaluated

(contd....)

Plant Name	*Month of Flowering and fruiting*	*IUCN Red List Category*
Calacanthus grandiflorus (Dalz.) Radlk.	Oct. – Mar.	Low Risk
Carvia callosa (Nees) Bremek.	Sept. – Jun.	Low Risk
Cynarospermum asperrimum (Nees) Vollesen	Oct. – Mar.	Low Risk
Dicliptera cuneata Nees	Dec. – Mar.	Data Deficient
**D. ghatica* Sant.	Apr. – Jun.	Data Deficient
**D. nasikensis* Lakmi. & Sharma	Mar. – Jun.	Endangered
Dyschoriste dalzellii (T. And. ex Bedd.) Ktze.	Sept. – Feb.	Low Risk
D. vagans (Wight) Ktze.	Sept. – Mar.	Not Evaluated
Eranthemum capense L. var. *concanense* (T. And. ex Cl.) Sant.	Oct. – Mar.	Data Deficient
E. roseum (Vahl.) R.Br.	Oct. – Mar.	Low Risk
Gantelbua urens (Heyne ex Roth) Bremek.	Dec. – Mar.	Low Risk
Gymnostachyum latifolium (Dalz.) T. And.	Nov. – Mar.	Not Evaluated
Haplanthodes nilgherrensis (Wight) Majumdar	Dec. – Jun.	Low Risk
H. plumosa (T. And.) Panigr. & Das	Jan. – Jun.	Low Risk
H. tentaculatus (L.) Majumdar	Dec. – May	Low Risk
H. verticillatus (Roxb.) Majumdar	Dec. – Jun.	Low Risk
Hygrophilla pinnatifida (Dalz.) Sreem.	Jan. – May	Low Risk
**Hypoestes lanata* Dalz.	Oct. – Feb.	Critically Endangered (?)
Indoneesiella echioides (L.) Sreem.	Jun. – Dec.	Low Risk
Justicia santapaui Bennet	Jan. – Feb.	Low Risk
J. trinervia Vahl	Oct. – Mar.	Low Risk
J. wynaadensis (Nees) T. And	Nov. – Mar.	Low Risk
**Lepidagathis bandraemis* Blatt.	Nov. – Mar.	Critically Endangered
L. lutea Dalz.	Nov. – Mar	Not Evaluated
L. mitis Dalz.	Nov. – Mar.	Low Risk
Mackenziea integrifolia (Dalz.) Bremek.	Dec. – May	Low Risk
Neuracanthus sphaerostachyus (Nees) Dalz.	Sept. – Jun.	Low Risk
Nilgirianthus asper Sant.	Dec. – May	Not Evaluated
N. barbatus (Nees) Bremek.	Oct. – Mar.	Low Risk
N. ciliatus (Nees) Bremek.	Oct. – Nov.	Endangered
N. heyneanus (Nees) Bremek.	Nov. – Mar.	Low Risk
N. lupulinus (Nees) Bremek.	Oct. – May	Low Risk
N. membranaceus (Talb.) Bremek.	Nov. – Feb.	Vulnerable
**N. reticulatus* (Stapf) Bremek.	Oct. – Mar.	Low Risk

(contd....)

Plant Name	*Month of Flowering and fruiting*	*IUCN Red List Category*
N. warrenensis (Dalz.) Bremek.	Oct. – Apr.	Endangered (?)
**Pleocaulis ritchiei* (Cl.) Bremek.	Sept. – Mar.	Low Risk
Rungia crenata T. And.	Sept. – Feb.	Data Deficient
R. linifolia Nees	Oct. – Feb.	Critically Endangered
Supushpa scrobiculata (Dalz. ex Cl.) Suryan.	Nov. - Mar.	Vulnerable
**Synnema anomalum* (Blatt.) Sant.	Oct. – Dec.	Data Deficient
Thelepaepale ixiocephala (Benth.) Bremek.	Nov. – Mar.	Low Risk
LAMIACEAE		
Anisochilus verticillatus Hook. f.	Aug. – Oct.	Not Evaluated
Anisomeles heyneana Benth.	Oct. - Mar.	Not Evaluated
Lavandula lawii Wight	Oct. - Mar.	Low Risk
**Leucas deodikarli* Billore & Hemadri	Sept. – Dec.	Endangered
**Nepeta bombainsis* Dalz.	Aug. – Oct.	Not Evaluated
Pogostemon deccanensis (Panigr.) Press	Sept. – Nov.	Not Evaluated
P. erectus (Dalz.) Ktze.	Oct. – Jan.	Not Evaluated
P. mollis Benth.	Not known	Data Deficient
P. myosuriodes (Benth.) Ktze.	Nov. – Jan.	Not Evaluated
P. purpurascens Dalz.	Oct. - May	Not Evaluated
P. salicifolius (Dalz. ex Hook. f.) Gazzar & Watson	Jan. – Apr.	Endangered
P. stocksii (Hook. f.) Press	Dec. – Mar.	Not Evaluated
AMARANTHACEAE		
**Achyranthes coynei* Sant.	Throughout the year	Data Deficient
Amaranthus caturus Heyne ex Hook.	Aug. – Nov.	Not Evaluated
PODOSTEMACEAE		
Cladopus hookeriana (Tul.) Cusset	Oct. – Feb.	Not Evaluated
Dalzellia zeylanica (Gardner) Wight	Sept. – Mar.	Not Evaluated
Hydrobryopsis sessilis (Willis) Engler	Oct. – Dec.	Not Evaluated
Polypleurum dichotomum (Gardner) Hall	Nov. – Jan.	Not Evaluated
P. stylosum (Wight) J.B.Hall	Oct. – Nov.	Not Evaluated
PIPERACEAE		
Peperomia hookeri Miq.	Apr. – Sept.	Low Risk
Piper trichostachyon (Miq.) Cl.	Aug. – Apr.	Low Risk

(contd....)

Plant Name	*Month of Flowering and fruiting*	*IUCN Red List Category*
MYRISTICACEAE		
Knema attenuata (Hook.f. et Thoms.) Warb.	Oct. – Jun.	Low Risk
Myristica malabarica Lamk.	Nov. – Mar.	Low Risk
LORANTHACEAE		
Dendrophthoe trigona (Wight & Arn.) Danser ex Sant.	Nov. – Mar.	Low Risk
Helicanthes elastica (Desr.) Danser	Aug. – Feb.	Low Risk
Helixanthera obtusata (Schult.) Danser	Feb. – Jun.	Not Evaluated
H. wallichiana (Schult.) Danser	Mar. – Jul.	Not Evaluated
**Scurrula stocksii* (Hook. f.) Dancer	Nov. – Mar.	Vulnerable
Tolypanthes lagenifer (Wight) van Tiegh.	Jun. – Sept.	Vulnerable
EUPHORBIACEAE		
Blachia denudata Benth.	Sept. – Mar.	Low Risk
Bridelia crenulata Roxb.	June	Data Deficient
Cleistanthus malabaricus Muell.-Arg.	Nov. – Jan.	Not Evaluated
Croton gibsonianus Nimmo	Nov. – Dec.	Data Deficient
Dimorphocalyx lawianus (Muell. – Arg.) Hook. f.	Oct. – Jun.	Low Risk
Drypetes venusta (Wight) Pax. & Hoffm.	Nov. – Apr.	Data Deficient
Epiprinus mallotiformis (Muell.-Arg.) Crozat	Oct. – Dec.	Data Deficient
Euphorbia concanensis Janarthanam & Yadav	Aug. – Oct.	Low Risk
E. fujiformis Buch. Ham. ex D. Don	Jan. – Mar.	Low Risk
**E. katrajensis* Gage	Sept. – Oct.	Data Deficient
**E. khandalensis* Blatt. & Hallb.	Jan. – May	Data Deficient
E. notoptera Boiss.	Apr. – Aug.	Low Risk
**E. panchganiensis* Blatt. & McC.	Feb. – Sept.	Data Deficient
E. sebastinei Binojkumar & Balakrishnan	Throughout	Not Evaluated
Glochidion ellipticum Wight	Feb. – Nov.	Low Risk
G. johnstonei	Feb. – Oct.	Data Deficient
G. malabaricum Bedd.	Jan. – May	Not Evaluated
G. tomentosum Dalz.	Feb. – Dec.	Data Deficient
**Jatropha nana* Dalz.	Jul. – Sept.	Vulnerable
Mallotus auro-punctatus Hook. f.	Oct. – May	Low Risk
M. stenanthus Muell.-Arg.	Oct. – Jun.	Not Evaluated
Phyllanthus scabrifolius Hook f.	September	Data Deficient
Trewia polycarpa Benth.	Jan. – Jun.	Not Evaluated

(contd....)

Plant Name	*Month of Flowering and fruiting*	*IUCN Red List Category*
URTICACEAE		
Pouzolzia integrifolia Dalz.	Aug. – Oct.	Low Risk
MONOCOTYLEDONS		
ORCHIDACEAE		
Aerides dalzelliana (Sant.) Garay	May – Mar.	Low Risk
A. maculosum Lindl.	May – Oct.	Low Risk
Bulbophyllum fimbriatum (Lindl.) Reichb.f.	Jun. – Apr.	Not Evaluated
B. neilgherrense Wight	Jul. – Dec.	Not Evaluated
Dendrobium aqueum Lindl.	Sept. - May	Low Risk
D. barbatulum Lindl.	Jan. – Jul.	Low Risk
D. herbaceum Lindl.	Sept. – Apr.	Low Risk
D. lawianum Lindl.	Mar. – Jul.	Low Risk
D. microbulbon A.Rich.	Dec. – May	Low Risk
D. nanum Gammie	Jul. – Apr.	Not Evaluated
D. ovatum (Willd.) Kranzl.	Sept. – Mar.	Not Evaluated
Diplocentrum recurvum Lindl.	May – Jul.	Data Deficient
Eria dalzelli Lindl.	Jul. – Oct.	Low Risk
E. exilis Hook.f.	Oct. – Mar.	Not Evaluated
E. microchilos (Dalz.) Lindl.	Jul. – Oct.	Low Risk
E. mysorensis Lindl.	Jul. – Oct.	Data Deficient
Eulophia ochreata Lindl.	Jun. – Jul.	Vulnerable
E. ramentacea Lindl.ex Wight	Dec. – Mar.	Not Evaluated
Flickingeria macraei (Lindl.) Seid.	Jul. – Oct.	Vulnerable
**Habenaria caranjensis* Dalz.		Data Deficient
H. crassifolia A.Rich.	Aug. – Oct.	Low Risk
**H. foliosa* A.Rich. var. *foliosa*	Jul. – Sept.	Low Risk
H. foliosa A.Rich. var. *foetida* (Blatt. & McC.) Bennet.	Jul. – Sept.	Low Risk
H. foliosa A.Rich. var. *gibsonii* (Hook.f.) Bennet	Jul. – Sept.	Vulnerable
H. grandifloriformis Blatt. & McC.	Jul. – Sept.	Low Risk
H. heyneana Lindl.	Aug. – Sept.	Low Risk
H. longicorniculata Grah.	Jul. – Oct.	Low Risk
H. multicaudata Sedgew.	Aug. – Sept.	Vulnerable
**H. ovalifolia* Wight	Aug. – Nov.	Low Risk
**H. panchganiensis* Sant. & Kap.	Aug. – Nov.	Vulnerable
H. rariflora A.Rich.	Aug. – Oct.	Low Risk
H. roxburghii Nicols.	Jul. – Oct.	Vulnerable
**H. suaveolens* Dalz.	Aug. – Sept.	Data Deficient
Loxoma maculata (Dalz.) Garay	May	Data Deficient
Oberonia brunoniana Wight	Feb. – Mar.	Low Risk
O. platycaulon Wight	Not Known	Data Deficient
P. stocksii (Hook.f.) Kranzl.	Jul. – Oct.	Low Risk

(contd....)

Plant Name	*Month of Flowering and fruiting*	*IUCN Red List Category*
Porpax jerdoniana (Wight) Rolfe	Jun. – Jul.	Vulnerable
P. reticulata Lindl.	Jun. – Oct.	Low Risk
Trias stocksii Benth. ex Hook. f.	Feb. – Nov.	Not Evaluated
ZINGIBERACEAE		
Curcuma amada Roxb.	Jun. – Sept.	Data Deficient
C. decipiens Dalz.	Jun. – Sept.	Low Risk
**C. inodora* Blatt.	Jun. – Sept.	Not Evaluated
C. neilgherrensis Wight	May – Jun.	Low Risk
C. pseudomontana Grah.	May – Sept.	Low Risk
**C. purpurea* Blatt.	Jun. – Jul.	Data Deficient
**Hitchenia caulina* (Grah.) Baker	Jul. – Sept.	Low Risk
Zingiber cernuum Dalz.	Jul. – Oct.	Low risk
Z. neesanum (Grah.) Ramam.	Jul. – Sept.	Low Risk
MUSACEAE		
Ensete superbum (Roxb.) Cheesm.	Jul. – Jan.	Low Risk
AMARYLLIDACEAE		
**Crinum brachynema* Herb.	Apr. – May	Critically Endangered
**C. elenorae* Blatt. & McC. f. var. *elenorae*	Apr. – May	Extinct (?)
**C. elenorae* Blatt. & McC. f. var. *purpurea* Blatt. & McC.	Apr. – May	Extinct (?)
**C. woodrowii* Baker	May – Jun.	Extinct (?)
Pancratium parvum Dalz.	Jun. – Aug.	Not Evaluated
**P. sanctae-mariae* Blatt. & Hallb.	May – Jun.	Critically Endangered
DIOSCORIACEAE		
Dioscorea belophylla (Prain) Haines	Aug. – Sept.	Low Risk
LILIACEAE		
Asparagus laevissimus Steud. ex Baker	Jul. – Aug.	Not Evaluated
**Camptorrhiza indica* Yadav, Singh & Mathew	Jun. – Jul.	Vulnerable
Chlorophytum bharuchae Ansari, Raghavan & Hemadri	May – Aug.	Vulnerable
C. borivilianum Sant. & Fernand.	May – Aug.	Low Risk
**C. glaucoides* Blatt.	Aug. – Nov.	Low Risk
C. glaucum Dalz.	Aug. – Nov.	Low Risk
**Dipcadi concanense* (Dalz.) Baker	Aug. – Sept.	Vulnerable
**D. maharashtrensis* Deb & Dasgupta	Aug. – Sept.	Data Deficient
**D. minor* Hook.f.	Not Known	Data Deficient
**D. saxorum* Blatt.	Jun. – Nov.	Vulnerable

(contd....)

Plant Name	*Month of Flowering and fruiting*	*IUCN Red List Category*
**D. ursulae* Blatt. var. *ursulae*	Aug. - Sept	Low Risk
D. ursulae Blatt. var. *longiracemosae* Deb. & Dasg.	Aug. - Sept.	Low Risk
Drimia polyantha (Blatt. & McC.) Ansari & Raghavan	Mar. - Jul.	Data Deficient
D. polyphylla. (Hook. f.) Ansari & Raghavan	Not Known	Data Deficient
D. razii Ansari	Mar. - Apr.	Vulnerable
Iphigenia magnifica Ansari & Rao	Aug. - Oct.	Low Risk
I. pallida Baker	Aug. - Oct.	Low Risk
**I. stellata* Blatt.	Jun. - Sept.	Low Risk
Protasparagus asiaticus Kamble	May - Aug.	Data Deficient
**Scilla viridis* Blatt. & Hallb.	Sept. Mar.	Data Deficient
COMMELINACEAE		
Belosynapsis vivipara (Dalz.) Sprague ex Fischer	Aug. - Sept.	Data Deficient
Commelina attenuata Koen. ex Vahl.	Aug. - Nov.	Low Risk
C. hirsuta (Wight) Cl.	Jul. - Nov.	Not Evaluated
Cyanotis arcotensis Rao	October	Not Evaluated
C. cerifolia Rao & Kammathy	Aug. - Nov.	Not Evaluated
C. concanensis Hassk.	Aug. - Nov.	Low Risk
C. fasciculata (Heyne ex Roth) Schult.var.*glabrescens* Cl.	Aug. - Oct.	Not Evaluated
C. papilionacea (L.) R. & S. var. *vaginata* Wight	Aug. - Oct.	Endangered
Murdania crocea (Griff.) Faden ssp. *ochracea* (Daz.) Faden	October	Not Evaluated
M. lanuginosa (Wall. ex Cl.) Brueck	May - Oct.	Endangered
M. versicolor (Dalz.) Brueck	Aug. - Oct.	Low Risk
Tonningia cucullata (Roth.) Ktze.	Jul. - Dec.	Not Evaluated
ARECACEAE		
Arenga wightii Griff.	Nov. - Jan.	Not Evaluated
Calamus thwaitesii Becc. & Hook. f.	Feb. - May	Not Evaluated
Hyphaene dichotoma (White) Furtado	Mar. - Sept.	Not Evaluated
**Phoenix robusta* Hook. f.	Jan. - Mar.	Vulnerable
ARACEAE		
Amorphophallus commutatus (Schott) Engl.	Apr. - Jul	Low Risk
A. konkanensis Hett., Yadav & Patil	Apr. - Jul.	Low Risk
**Arisaema caudatum* Engl.	Jul. - Oct.	Low Risk
**A. murrayii* (Grah.) Hook.f.	Jul. - Oct.	Vulnerable
**A. sahyadricum* Yadav, Patil & Bachulkar	Jul. - Sept.	Vulnerable

(contd....)

Plant Name	*Month of Flowering and fruiting*	*IUCN Red List Category*
**A. sivadasanii* Yadav, Patil & Janarthanam	Sept. – Oct.	Vulnerable
**Cryptocoryne cognata* Schott	Oct. – Jun.	Low Risk
**C. spiralis* (Retz.) Fisch. ex Wydl. var. *cognatoides* (Blatt. & McC.) Yadav, Patil & Bogner	Oct. – Jun.	Low Risk
Theriophonum dalzellii Schott	Jun. – Sept.	Low Risk
Typhonium bulbiferum Dalz.	Jun. – Jul.	Not Evaluated
ALISMATACEAE		
Wiesneria triandra (Dalz.) Mich.	Jul. – Sept.	Low Risk
APONOGETONACEAE		
**Aponogeton bruggenii* Yadav & Govekar	Jun. – Aug.	Critically Endangered
**A. satarensis* Raghavan, Yadav & Kulkarni	Jun. – Sept.	Vulnerable
HYDATELLACEAE		
Trithuria konkanensis Yadav & Janarthanam	Jul. – Sept.	Low Risk
ERIOCAULACEAE		
Eriocaulon breviscapum Koern.	Oct. – Dec.	Vulnerable
E. conica (Fyson) Fischer	Aug. – Oct.	Low Risk
E. cuspidatum Dalz.	Aug. – Oct.	Low Risk
E. dalzellii Koern.	Aug. – Oct.	Low Risk
E. duthiei Hook. f.	Sept. – Oct.	Low Risk
E. elenorae Fyson	Aug. – Nov.	Low Risk
E. eurypeplon Koern.	Aug. – Oct.	Low Risk
E. fluviatile Trimen	Oct. – Dec.	Vulnerable
E. heterolepis Steud.	Aug. – Oct.	Low Risk
**E. kolhapurense* Gaikwad, Sardesai & Yadav	Aug. – Oct.	Vulnerable
E. lanceolatum Miq. ex Steud.	Aug. – Oct.	Low Risk
E. lanceolatum Miq. ex Steud. var. *pilosum* Mold.	Aug. – Oct.	Low Risk
E. margaretae Fyson	Aug. – Nov.	Low Risk
E. martianum Wall. ex Koern.	Aug. – Nov.	Low Risk
E. minutum Hook. f.	Aug. – Nov.	Low Risk
E. odoratum Dalz.	Aug. – Oct.	Low Risk
E. palghatense Ans. & Balak.	5857	Low Risk
**E. ratnagiricus* Yadav, Gaikwad & Sardesai	Aug. – Sept.	Not Evaluated
E. reductum Ruhland	Aug. – Oct.	Low Risk
E. richardianum (Fyson) Ans. & Balak.	Aug – Sept.	Not Evaluated

(contd....)

Plant Name	*Month of Flowering and fruiting*	*IUCN Red List Category*
E. ritchieanum Ruhland	Aug. – Oct.	Low Risk
E. robusto-brownianum Ruhland	Aug. – Oct.	Low Risk
**E. santapaui* Moldenke	Aug. – Sept.	Data Deficient
E. sedgwickii Fyson	Aug. – Oct.	Low Risk
E. setaceum L.	Aug. – Sept.	Low Risk
**E. sharmae* Ans. & Balak.	Aug. – Sept.	Low Risk
E. stellulatum Koern.	Aug. – Sept.	Low Risk
**E. talbotii* Ansari & Balak.	Aug. – Sept.	Data Deficient
**E. tuberiferum* Kulkarni & Desai	Aug. – Sept.	Low Risk
CYPERACEAE		
Bulbostylis subspinescens Cl.	Jul. – Sept.	Not Evaluated
Carex caricina (D. Don) Ghildyal & Bhattacharyya *var. glaucina* (Boeck) Ghildyal & Bhattacharyya	Sept. – Mar.	Not Evaluated
**Cyperus decumbens* Govind.	Not Known	Data Deficient
**C. pentabracteatus* Govind & Hemadri	Sept. – Oct.	Endangered
**Eleocharis lankana* Koyama *ssp. mohamadi* WadoodKhan	Oct. – Dec.	Endangered
**Fimbrystylis ambavanensis* Prasad & Singh	Sept. – Oct.	Endangered
**F. nagpurensis* Prasad & Singh	August	Data Deficient
**F. ratnagirica* Prasad & Singh	August	Data Deficient
**F. unispicalaris* Govind. & Hemadri	Sept. – Oct.	Critically Endangered
**Mariscus blatteri* McC.	October	Critically Endangered
**M. konkanensis* (T.Cooke) Sedgw.	Jul. – Nov.	Low Risk
**Pycerus bolei* Almeida	Oct. – Nov.	Data Deficient
**P. lancelotii* Almeida	Oct. – Nov.	Data Deficient
**P. malabaricus* C.B.Cl.	Dec. – Jan.	Low Risk
**Scirpus naikianus* Wadood Khan	Dec. – Jan.	Critically Endangered
**S. poklii* Wadood Khan	Oct. – Nov.	Critically Endangered
POACEAE		
Aristida stocksii (Hook.f.) Domin.	Sept. – Nov.	Not Evaluated
Arthraxon hispidus (Thunb.) Makino var. *junnarensis* (Jain & Hemadri) Welzen	Sept. – Oct.	Endangered
A. hispidus (Thunb.) Makino var. *santapaui* (Bor) Welzen	Sept. – Oct.	Endangered
A. jubatus Hack.	Oct. – Feb.	Low Risk
A. lanceolatus (Roxb.) Hochst. var. *meeboldii* (Stapf) Welzen	Sept. – Oct.	Not Evaluated

(contd....)

Plant Name	*Month of Flowering and fruiting*	*IUCN Red List Category*
A. lanceolatus (Roxb.) Hochst. var. *raizadae* (Jain *et al,*) Welzen	Sept. – Oct.	Not Evaluated
A. lanceolatus (Roxb.) Hochst. var. *villosus* (Fisch.) Welzen.	Oct. – Nov.	Not Evaluated
Arundinella leptochloa (Nees ex Steud.) Hook. f.	Oct. – Jun.	Not Evaluated
A. metzii Hochst. ex Miq.	Oct. – Dec.	Not Evaluated
A. nervosa (Roxb.) Nees ex Hook. & Arn.	Oct. – Nov.	Not Evaluated
A. purpurea Hochst. ex Steud. var. *purpurea*	Oct. – Nov.	Not Evaluated
A. spicata Dalz.	Sept. – Nov.	Low Risk
A. tuberculata Munro ex Lisboa	Sept. – Nov.	Not Evaluated
Bhidea burnsiana Bor	Aug. – Dec.	Low Risk
Brachiaria semiundulata (Hochst.ex A. Rich) Stapf	Sept. – Oct.	Not Evaluated
Chloris bournei Rang. & Tad.	Sept. – Jan.	Not Evaluated
C. quinqiusetica Bhide	Aug. – Jan.	Not Evaluated
Chrysopogon asper (Heye ex Hook.f.) Blatt. & McC.	Sept. – Oct.	Not Evaluated
**C. castneus* Veldkamp & Salunke	Sept. – Oct.	Not Evaluated
Chrysopogon hackelii (Hook. f.) Fisch.	Aug. – Nov	Not Evaluated
C. orientalis (Desv.) A. Camus	Sept. – Oct.	Not Evaluated
Cleistachne stocksii Hook. f.	Oct. – Dec.	Not Evaluated
**Coelachne minuta* Bor	Aug. – Nov.	Vulnerable
Cynodon barberi Rang. & Tad.	Aug. – Nov.	Not Evaluated
**Danthonidium gammiei* (Bhide) Hubb.	Sept. – Oct.	Vulnerable
**Dichanthium armatum* (Hook. f.) Blatt. & McC.	Sept. – Oct.	Vulnerable
**D. compressum* (Hook.f.) Jain & Deshpande	Oct. – Nov.	Endangered
D. concanensis (Hook.f.) Jain & Deshpande	Oct. – Nov.	Not Evaluated
D. filiculme (Hook. f.) Jain & Deshpande	Sept. – Nov.	Not Evaluated
**D. jainii* (Deshpande & Hemadri) Deshpande	Oct. – Jan.	Endangered
**D. maccanii* Blatt.	Oct. – Apr.	Data Deficient
D. odoratum (Lisboa) Jain & Deshpande	Oct. – Dec.	Not Evaluated
D. oliganthum (Hochst. & Steud.) T.A.Cope	Oct. – Nov.	Not Evaluated
**D. panchganiensis* Blatt. & McC.	Oct. – Nov.	Endangered
D. paranjpyeanum (Bhide) Clayton	Oct. – Nov.	Not Evaluated

(contd....)

Plant Name	*Month of Flowering and fruiting*	*IUCN Red List Category*
D. tuberculatum (Hack.) T.A. Cope	Oct. – Nov.	Not Evaluated
**D. woodrowii* (Hook. f.) Jain & Deshpande	Dec. – Jan.	Endangered
**Dimeria blatteri* Bor	Oct. – Nov.	Vulnerable
D. deccanensis Bor	Oct. – Dec.	Not Evaluated
D. hohenackeri Hochst. ex Miq.	Oct. – Feb.	Not Evaluated
D. ornithopoda Trin var. *megalantha* Bor	Oct. – Nov.	Not Evaluated
D. santapaui Almeida	Sept. – Oct.	Not Evaluated
D. stafiana Hubb.ex Pilger	Sept. – Feb	
D. woodrowii Stapf	Sept. – Feb	Vulnerable
Garnotia arborum Stapf ex Cooke	Sept. – Dec.	Not Evaluated
Glyphochloa acuminata (Hack.) Clayton var. *acuminata*	Aug. – Feb.	Not Evaluated
G. acuminata (Hack.) Clayton var. *stocksii* (Hook.f.) Clayton	Sept. – Oct.	Not Evaluated
G. acuminata (Hack.) Clayton var. *woodrowii* (Bor) Clayton	Oct. – Nov.	Not Evaluated
G. divergens (Hack.) Clayton var. *divergens*	Sept. – Oct.	Vulnerable
G. divergens (Hack.) Clayton var. *hirsuta* (Fisch.) Clayton	Sept. – Oct.	Not Evaluated
G. forficulata (Fisch.) Clayton	Sept. – Nov.	Not Evaluated
G. goaensis (Rao & Hemadri) Clayton	Oct. – Nov.	Not Evaluated
G. mysorensis (Jain & Hemadri) Clayton	Sept. – Oct.	Not Evaluated
G. ratnagirica (Kulkarni & Hemadri) Clayton	Sept. – Oct.	Not Evaluated
**G. santapaui* (Jain & Desh.) Clayton	Aug. – Feb.	Vulnerable
G. talbotii (Hook.f.) Clayton	Sept. – Oct	Vulnerable
**Heteropogon polystachyos* (Roxb.) Schult.	Jul. – Nov.	Not Evaluated
**H. ritchiei* (Hook.f.) Blatt. & McC.	Sept. – Nov.	Vulnerable
Hubbardia haptaneuron Bor	Aug. – Sept.	Critically Endangered
Indopoa paupercula (Stapf) Bor	Aug. – Nov.	Not Evaluated
Isachne angladei Fisch.	Oct. - Nov.	Not Evaluated
**I. bicolor* Naik & Patunkar	Aug. – Dec.	Not Evaluated
**I. borii* Hemadri	Aug. – Dec.	Vulnerable
I. elegans Dalz.	Sept. – Jan.	Not Evaluated
I. gracilis Hubb.	Sept. – Dec.	Not Evaluated
I. lisboae Hook.f.	Sept. – Oct.	Not Evaluated
I. meeboldii Fisch.	Aug. – Dec.	Not Evaluated
I. swaminathanii Ved Prakash & Jain	Sept. – Oct.	Not Evaluated

(contd....)

Plant Name	*Month of Flowering and fruiting*	*IUCN Red List Category*
**Ischaemum bolei* Almeida	Oct. – Nov.	Data Deficient
**I. bombaiense* Bor	Oct. – Nov.	Not Evaluated
I. dalzellii Stapf ex Bor	Oct. – Nov.	Not Evaluated
**I. diplopogon* Hook. f.	Sept. – Nov.	Not Evaluated
**I. huegellii* Hack.	Not Known	Data Deficient
**I. impressum* Hack.	Sept. – Jan.	Not Evaluated
I. molle Hook.f.	Oct. – Nov.	Not Evaluated
I. pilosum (Klein ex Willd.) Wight	Sept. – Mar.	Not Evaluated
I. raizadae Hemadri & Billore	Sept. – Nov.	Vulnerable
I. ritchiei Stapf ex Bor	Oct. – Nov.	Not Evaluated
**I. santapaui* Bor	Sept. – Nov.	Not Evaluated
I. thomsonianum Stapf ex Fisch.	Nov. – Dec.	Not Evaluated
I. travancorense Stapf ex Fisch.	Oct. – Feb.	Not Evaluated
I. tumidum Stapf ex Bor	Sept. – Dec.	Not Evaluated
Iseilema anthephoroides Hack.	Aug. – Sept.	Not Evaluated
Lophopogon tridentatus (Roxb.) Hack.	Aug. – Dec.	Low Risk
Mnesithea clarkei (Hack.) Koning & Sosef	Sept. – Nov.	Not Evaluated
**Ophiuros bombaiensis* Bor	Jul. – Oct.	Vulnerable
Oropetium roxburghianum (Steud.) Phillips	Aug. – Nov.	Low Risk
Panicum deccanenese Naik & Patunkar	Sept. – Oct.	Not Evaluated
P. johnii Almeida	Sept. – Oct.	Endangered
**P. painanum* Naik & Patunkar var. *paianum*	Sept. – Oct.	Vulnerable
**P. painanum* Naik & Patunkar var. *minor* Naik & Patunkar	Oct. – Nov.	Vulnerable
**P. phoiniclados* Naik & Patunkar	Oct. – Nov.	Critically Endangered
Paspalum canarae (Steud.) Veldk. var. *canarae*	Aug. – Oct.	Not Evaluated
**Pogonachne racemosa* Bor	Oct. – Nov.	Vulnerable
Pseudanthistiria hispida Hook. f.	Sept. – Nov.	Low Risk
P. intermedia Birari & D'Cruz	Nov. – Dec.	Not Evaluated
Pseudodichanthium serrafalcoides (Cooke & Stapf) Bor	Sept. – Nov.	Vulnerable
Schizachyrium paranjpyeanum (Bhide) Raizada & Jain	Sept. – Nov.	Endangered
Sehima sulcatum (Hack.) A. Camus	Aug. – Nov.	Not Evaluated
**Themeda pseudotrimula* Potdar, Salunkhe & Yadav	Sept. – Nov.	Endangered
Tragus roxburghii Panigrahi	Aug. – Sept.	Low Risk
Trilobachne cookei (Stapf) Schenck. & Henrard	Aug. – Oct.	Endangered

(contd....)

Plant Name	Month of Flowering and fruiting	IUCN Red List Category
**Triplopogon ramosissimus* (Hack.) Bor	Aug. - Oct.	Endangered
Tripogon bromoides R. & S.	Jul. - Dec.	Not Evaluated
T. capillatus Jaub. & Spach.	Aug. - Nov.	Not Evaluated
T. jacquemontii Stapf	Aug. - Nov.	Not Evaluated
**T. polyanthus* Naik & Patunkar	Aug. - Dec.	Endangered
T. pungens Fisch.	Sept. - Nov.	Not Evaluated
Urochloa panicoides P. Beauv. var. *marathensis* (Henr.) Bor	May - Sept.	Low Risk
Vetiveria lawsonii (Hook.f.) Blatt. & McC.	Oct. - Dec.	Not Evaluated

REFERENCES

Ahmedullah, M. and Nayar, M.P. 1986. *Endemic plants of the Indian region* **vol. 1**. Botanical Survey of India, Calcutta.

Blasco, F. Montagnes. Du Sud De L'nde. 1971. Inst. For. Pondichury Trans. Sec. Sci. and Tech. Vol. 10.

Bramwell, D. 1972. Endemism in the flora of the Canary Island, in Valentine, D.H. (ed.) *Taxonomy, Phytogeography and Evolution*, pp. 141–159. Acad. Press. London.

Briquet, J. 1905. *Les reimmigrations postglaciaires des flores en Suisse*. Act. Soc. Helv. Sci. Nat. Vol. 1.

Chatterjee, D. 1940. Studies of the endemic flora of India and Burma, *J. Asiat. Soc. Bengal, Sci.* **5 (1)**: 19–68.

Chevalier, A and Guenot, L. 1925. *Biogeographie Chap. X. Pueplement des espaces vides*. Paris.

Cooke, T. 1901–1908. *The Flora of the Presidency of Bombay*. London. 2. Vols. (Repr. ed. 1958. 3 Vols. Govt. of India).

Decandolle, A.P. 1855. *Geographe Botanique Raisonnee* Vols. 1 and 2. Geneva.

Engler, A. 1882. Versuch einer Entwick Lungsgeschichte der pflagenwelt Leipzig.

Diels, L. 1909. *Geographie der Pflanzengeophia*. Stuttgart.

Drude, O. 1890. *Handbuch der Pflazengeographie*. Stuttgart.

Herzog, Th. 1926. *Geographie der Moose* Jena.

Jain, S.K. and Rao, R.R. (eds.). 1983. *An Assessemnt of Threatened Plants of India*, Proceeding of the seminar held at Dehradun, 14th–17th Sept. 1981.

Mishra, D. and Singh, N.P. 2001. *Endemic and Threatened Flowering Plants of Maharashtra*. Botanical Survey of India. Calcatta.

Mistry, K.M. and Almeida, S.M. 1989. Some rare, endangered and threatened plant species from Ratnagiri district, Maharashtra, *J. Bombay Nat. Hist. Soc.* **86(3)**: 478–479.

Nayar, M.P. and A.R.K. Sastry (eds.). 1987, 1988 and 1990. *Red Data Book of Indian Plants*, Vols. **1**, **2** and **3**. Botanical Survey of India, Calcatta.

Nayar, M.P. 1977. Changing Patterns of the Indian Floras, *Bull. Bot. Surv. India* **19**: 145–155.

Nayar, M.P. 1996. *Hot Spots of Endemic plants of India, Nepal and Bhutan*. Tropical Botanic Garden and Research Institute, Palode, Thiruvananthapuram.

Raghavan, R.S. and Singh, N.P. 1983. in S.K. Jain and A.R.K. Sastry (eds.). Plant Conservation Bulletin, Vol. 3. POSSCEF, Botanical Survey of India. Calcatta.

Raghavan, R.S. and Singh, N.P. 1984. An inventory of endemic and vulnerable species of Western India deserving conservation, *J. Econ. Tax. Bot.* **5(1)**: 153–164.

Rao, C.K. 1972. Angiosperm genera endemic to the Indian floristic region and its neibouring areas, *Indian For.* **98**: 560–566.

Richardson, I.B.K. 1978. Endemic Taxa and the Taxonomist, in Street, H.E. (ed.). Essays in Plant Taxonomy, pp. 245–262. Acad. Press.

Singh, N.P. and Raghavan, R.S. 1986. Materials for plant conservation in Western India, *J. Econ. Tax. Bot.* **8(1)**: 29–38.

Srinivasan, K. 1959. Protection of wild plant life. *ibid.* **1**: 6–10.

Subramanyam, K. and M.P. Nayar. 1974. Vegetation and Phytogeography of the Western Ghats in M.S. Mani (ed.) *Ecology* and *Biogeography in India.* The Hague, Netherlands.

Subramanyam, K. and Sreemadhavan C.P. 1969. Endangered plant species and their habitats — a review of the Indian situation. 11th Tech. Meeting of IUCN. New Delhi.

Vajravelu, E. and Daniel, P. 1983. in S.K. Jain and A.R.K. Sastry (eds.). *Materials for a catelogue of threatened plants of India.* Botanical Survey of India, Calcatta.

Wulff, E.V. 1950. *An Introduction to Historical Plant Geography.* I. 223. Walthan, Massachussetts, U.S.A.

Yadav, S.R. 1997a. Endemic plants of Peninsular India with special reference to Maharashtra; in Pokle D.S.; Nanir, S.P. Naik, V.N. (eds.), Proceeding, VII IAAT Annul Meeting and National Conference. pp. 31–51, Aurangabad.

Yadav, S.R. 1997b. Rare Flowering plant species of Maharashtra, their potential values, utilization and conservation in sustainable development. Proc. Nat. Conf. Dimens. Environ. Stress India, Geol. Deptt. M. S. Univ. Baroda.

CHAPTER 4

STUDIES ON BLACK GRAM [*VIGNA MUNGO* (L.) HEPPER] AND *RHIZOBIUM* ASSOCIATION — A REVIEW

*K.V. Mallaiah**

Introduction

Black gram [*Vigna mungo* (L.) Hepper] is an economically important pulse crop, cultivated throughout the Asian Continent, some African countries and America, under a wide variety of edaphic and climatic conditions. In South-East Asian countries, it is often the most prized of all grain legumes (Kay, 1979). It is thought to have originated in India (De Condolle, 1886), and references to the crop were present in vedic texts, Kautilya's 'Arthasastra', and also in 'Materia medica' and 'Charaka Samhita' (Kachroo, 1970) indicating the importance and long history of the crop in India. It is popularly known as Urid or Urad or Urd bean in North India, as Mash in Bengal, 'Minumulu' in Andhra Pradesh and 'Ulundu' in Tamilnadu. In West Indies, it is called 'wolly pyrol'. It is rich in protein and phosphoric acid, and is used extensively in preparation of highly delicious and nutritious dishes like 'Idli', 'Vada', 'Dosa' and also in preparation of 'dhal' and a number of sweets.

The plant belongs to the family Fabaceae (Leguminosae), and being a leguminous crop possessing root nodules, it fixes atmospheric nitrogen and improve soil fertility. Hence, it is commonly cultivated as mixed crop subsidiary to cotton and millets, and as a second crop after

* Department of Botany and Microbiology, Nagarjuna University, Nagarjunanagar 522 510. Guntur Dist., A.P., India

the harvest of paddy both in rainy and winter seasons. It is also cultivated as early season rain-fed sole crop in upland areas. The plant is a fast growing erect herb up to 0.8 m in height with large trifoliate leaves. Flowering starts by 30th day after sowing and reaches peak between 45–50 days. The flowers are small yellow and typically papilionaceous. The pods are 4 to 6 cm in length, round, erect with 4–10 seeds per pod. The plant matures in 60-90 days depending on the cultivar. The yield of dried seeds range from 340 to 500 kg/ha, and the yield of dry forage usually weigh about three times as much (Duke, 1981).

Since the classical work of Hellriegal and Wilfarth (1888) which conclusively proved that nodules on the roots of leguminous plants are formed due to bacterial infection, and bacteria in the nodules fix atmospheric nitrogen, much work has been carried out on symbiotic association of bacteria with legume crops including black gram.

Root Nodule Bacterium of Black Gram

a) Identity of the Symbiont

The microsymbiont associated with the root nodules of black gram is treated for a long time as *Rhizobium* sp. belonging to 'cow pea miscellany' group, following the classical studies of Fred *et al.* (1932) on cross inoculation grouping of legume rhizobia. Raju (1936) conducted extensive cross inoculation studies in India and placed the root nodule bacterium of black gram in 'cow pea group', as it was distinctly different from bean group rhizobia in its cross inoculation capabilities. Dadarwal *et al.* (1977) reported that black gram *Rhizobium* shows strong serological affinities to the rhizobia from green gram, red gram and groundnut, all belonging to 'cow pea miscellany group'.

Even earlier, when black gram was placed in the genus *Phaseolus* as *P. mungo,* the root nodule bacterium of black gram was not placed in 'bean group' because it did not show cross infective ability with temperate species of *Phaseolus,* but showed free cross inoculation with *Vigna* species and others of 'cow pea group'. The taxonomic revision of the genus *Phaseolus* and placing of tropical species of *Phaseolus,* including black gram, in the genus *Vigna* solved the anomaly of placing rhizobia infecting *Phaseolus* species in two separate cross inoculation grouping viz. 'bean group' and 'cow pea group'. As Allen and Allen (1981) remarked "rhizobia, here, proved to have a more discriminating taxonomic judgment than plant taxonomists and thus we have a fine example of how 'symbiotaxonomy' can provide valuable supporting evidence for phylogenetic kinships".

Jordan (1982) separated slow growing, non-acid producing rhizobia from fast growing, acid producing strains, and placed them in a separate genus *Bradyrhizobium*, and the soybean *Rhizobium* (*Rhizobium japonicum*) was made the type species of the new genus as *Bradyrhizobium japonicum*. Later, Jordan (1984) transferred the rhizobia of cow pea miscellany group to the genus *Bradyrhizobium* and suggested that they be treated as *Bradyrhizobium* sp. with host name in the parenthesis.

The placement of cow pea rhizobia in *Bradyrhizobium* has not been accepted by all Rhizobiologists, since cow pea rhizobia are not always slow growing strains producing alkali (Hernandez and Focht, 1984). As a result, there is some confusion in using the generic name for 'cow pea rhizobia'. Some treat them as *Bradyrhizobium* sp. and some treat them as *Rhizobium* sp. with host name in parenthesis.

Presently the taxonomy of *Rhizobium* is undergoing rapid changes with the introduction of molecular techniques of analysis, especially nucleotide sequencing of 16 S rRNA. The earlier concept that a single *Rhizobium* species nodulate a group of closely related host species is seriously questioned now. Young and Huakka (1996) pointed out "although most *Rhizobium* isolates can nodulate more than one host species, it is also true that several different bacterial species are often isolated from a single legume species" and "the number of *Rhizobium* species might even be similar to the number of legume species currently estimated at up to 18,000". Mpeperaki *et al.* (1996) reported that contrary to the belief that 'cow pea rhizobia' are generally promiscuous, the cow pea rhizobia indigenous to Zimbabwean soils have relatively narrow host range. Pueppke and Broughton (1999) reported that *Vigna* species are more effectively nodulated by *Rhizobium fredii* than even rhizobia of cow pea miscellany. Herrera-Cervera *et al.* (1999) reported that atleast five rhizobial species nodulate *Phaseolus vulgaris* in Spanish soils. Hence, it is convenient to treat the microbial symbiont that nodulate black gram as *Rhizobium* sp. (black gram), until it is more properly characterized.

b) Cultural Characters of the Microbial Symbiont

The microbial symbiont of black gram root nodules is commonly isolated and maintained on Yeast Extract Mannitol Agar (YEMA) medium, like other rhizobia. Studies on cultural characters of black gram *Rhizobium* showed that it is a typical *Rhizobium*. The colonies are white, round to irregular, mucilaginous, slightly elevated in the middle and smooth in texture. Gupta and Sen (1963) reported that utilization of asparagine as nitrogen source, and glucose as carbon source is significantly correlated to its effectiveness. Optimum pH for growth is 7.0 (Gandhi and Vyas, 1969). It is sensitive to chloride and sulphate salts of sodium and potassium (Yadav and Vyas, 1971), but relatively more tolerant to salts than rhizobia from winter legumes (Gupta, 1969). It is sensitive to vanillin (Gupta *et al.*, 1974).

Juwarkar and Rewari (1988) studied the viability of *Rhizobium* at different temperatures and relative humidities, and reported that 50% RH at different storage temperatures extended the shelf life of the bacterium but higher RH in conjunction with high temperatures (40° and 45°C) resulted in rapid loss of viability.

Ahlawat *et al.* (1991) compared the growth of three isolates of *Rhizobium* sp. of cow pea miscellany group (S-24, CP 8 and G 20), and reported that under cultural conditions the strains were neutral without effecting the growth of one another. Ahlawat *et al.* (1992) reported that under cultural conditions the bacteriocin producing mutants of cowpea miscellany rhizobia inhibited the growth of standard strain of *Rhizobium* (S-24) having affinity to black gram and green gram.

Dadarwal and Sen (1973) reported that seed diffusates from various legumes, including black gram, showed inhibitory effect on the growth of *Rhizobium*. Jain and Rewari (1976) reported that the seeds of black gram showed inhibitory effect on *Rhizobium* isolates, and attributed it to diffusible toxic substances in the seed coat. It was further reported that presoaking of seeds in water decreased seed coat toxicity. Konde *et al.* (1982) studied the effect of seed coat diffusates of 15 black gram cultivars on growth of *Rhizobium*, and reported that under cultural conditions diffusates from 7 cultivars inhibited the growth of *Rhizobium*, diffusates of two cultivars stimulated the growth, while the remaining six cultivars had no effect on *Rhizobium*. Jain and Rewari (1974) reported that some bacteria and actinomycetes isolated from the seeds of black gram showed antagonistic activity to homologous *Rhizobium*, but fungal isolates showed no such antagonistic effect.

Nodulation Characteristics of Black Gram

a) Nodulation Pattern

The nodulation of black gram was widely reported by different workers (Allen and Allen, 1981). In India, Raju (1936) first studied the nodulation of black gram. The nodules are globose in shape, pink in colour, measure 3–5 mm in diameter and occur both as singles and in groups on both tap root and lateral roots (Reddy and Mallaiah, 2002).

Chakrabarty and Agarwal (1990) studied the nodulation pattern in 16 cultivars of black gram and reported that the cultivars can be divided into three groups viz. tap root nodulated, both tap root and lateral roots nodulated and lateral roots nodulated. They also reported that the cultivars within each group can be further subdivided on the basis of size of the nodules and the total number of nodules per plant.

The nodules appear within one week after sowing, increase in number with age up to 45 days when the plants are in full bloom, and as the plants mature they usually shrivel, become black and drop off. The number of nodules per plant vary widely from as low as 13 (Rangaswamy and Oblisami, 1962), 25–37 (Vaishya and Gajendragadkar, 1982), or 85 or more (Reddy and Mallaiah, 2002). The nodule fresh weight range from 75 to 101 mg and dry weight from 12 to 18 mg (Mathur *et al.*, 1998). Vaishya and Gajendragadkar (1982) reported that host genotype significantly effect the number of nodules per plant. Pawar and Ghulgule (1980) reported that maximum number of nodules occur by 21 days after sowing and senescence commence by 28 days, while Senaratne and Ratnasinghe (1993) reported that maximum nodulation occur during full bloom stage and senescence of nodules begin from mid pod filling stage. Ramaswamy *et al.* (1977) studied the dry weight of nodules and nitrogen content at different stages of plant growth, and reported that nitrogen content was high in nodules up to 60 days. Bushby (1984) reported that nodule decay was consistently associated with a large increase in the number of rhizobia per root system. Tuladhar (1997) reported that the degree of effective nodulation varied greatly among legumes and it was better in cow peas than in black gram.

b) Nodule Development

Lectins are host specific proteins that specifically bind to carbohydrate receptors on *Rhizobium* cells, and it is the first step in successful nodulation. Jayapragasam (1989) reported that lectin content in 26 cultivars of black gram differ significantly but there was no significant correlation between the amount of lectin and number of nodules formed per plant. Once infection occur successfully and nodulation begins, ploidy of the nodular tissue often changes. Schwent *et al.* (1983) reported that nodular tissue of black gram contain both diploid and tetraploid cells but no other degrees of ploidy. It was also reported that morphology of bacteroids in the nodules was only slightly different from living cultured bacteria in size and shape. Prasad (1982) studied the ultrastructure of black gram root nodule meristem and reported that mainly three types of tissues constitute the earliest detectable nodule, i.e., cortical, meristematic and central infected cells with feebly developed vascular elements. The nodule meristem develops in a crescent shaped manner resulting in spherical nodules with central infected bacteroid zone. Meristem cells appear as a brick-like layer in cross section. The nuclei are invariably large. As nodules develop, these meristem cells divide in both directions and contribute towards the formation of outer most layer of highly vacuolated nodule husk. The developing cells show a variety of interesting mitochondrial profiles. As the meristematic cells in a mature nodule differentiate, there is rapid disintegration of cytoplasmic bodies and peculiarities in mitochondrial morphology disappear. The nodule attains maximum size in about 15 days, and at this stage, the nodule meristem could hardly be distinguished. Prasad and Prasad (1989) reported that infected cells showed submicroscopic changes from the earliest detectable stage of infection and the changes in host tissue are associated with bacterial growth and activity. Selvaraj and Abragam (1999) reported that calcium oxalate crystals were present in large numbers in the outer, middle and inner regions of the nodule cortex. It is attributed to the positive preference of the plant for nitrate assimilation over nitrogen fixation.

c) Leghaemoglobin and Nodule Enzymes

Leghaemoglobin is a specific conjugated protein present in the legume root nodules and impart red colour to the nodular tissue. Its presence and amount is directly correlated to the effectiveness, i.e., ability to fix nitrogen, and amount of bacteroid tissue in the nodule (Singh, T.B. 1989). The leghaemoglobin content vary from 0.22 mg to 0.59 mg per nodule in different cultivars (Mathur *et al.*, 1998), and the content is more in nodules on plants grown in light than those grown in darkness (Saran, 1995). The leghaemoglobin content is also influenced by the *Rhizobium* strain that forms the nodule and use of phosphorus fertilizers. Dadarwal *et al.* (1982) reported that leghaemoglobin content of nodules formed by *Rhizobium* strain M-11 in cowpea miscellany hosts, including black gram, was 1.5 to 2 times more than that in nodules formed by *Rhizobium* strain S-24. Singh, M. *et al.* (1983) reported that use of phosphorus fertilizers significantly increase the dry weight and leghaemoglobin content of the nodules.

The nitrogenase enzyme is the most essential enzyme in nitrogen fixation, and its activity varies widely depending on the host and environmental factors which clearly reflect as seasonal and diurnal changes. The seasonal peak activity of nitrogenase enzyme occur during flowering stage in black gram (Shanmugam and Rangasamy, 1983) and diurnal peak occur at noon (Garg and Tauro, 1985). Shanmugam *et al.* (1984) reported that nodules showing higher nitrogenase activity at later stages of plant growth are highly desirable, and reported that out of the 9 genotypes screened, three showed high nitrogenase activity at the pod developing stage. Jaiswal *et al.* (1982) reported that the cytokinins viz. zeatin and zeatinribose, stimulated the nitrogenase activity in the nodules. Sengupta *et al.* (1986) reported that treatment of black gram seeds with disodium hydrogen phosphate and polyethylene glycol increased the nitrogenase activity and nitrogen fixation. Bose *et al.* (1987) reported that treatment of black gram seeds with leaf extracts of *Moringa oleifera* effectively increased the nitrogenase activity. Ramanujam *et al.* (1998) reported that pretreatment of black gram seeds with salicylic acid reduced the number of nodules formed, and the nitrogenase activity as swell as leghaemoglobin content also reduced. Lahiri *et al.* (1999) reported that treatment of excised root nodules of black gram with 0.05 to 0.1 mM spermine increased the activity of nitrogenase but higher concentration (up to 5mM) reduced the activity. They also reported that ribonuclease activity decreased with increasing spermine concentration while protease activity decreased sharply with 0.5 mM spermine. Hexokinase and Ü-amylase showed initial increase in activity with 0.5 and 0.1 mM spermine followed by decrease.

Chatterjee *et al.* (1983) reported that during nodule initiation and growth two peaks of polyamine content were noted, at 2 and 5 weeks after nodulation. In the roots decarboxylase activity increased up to 2 weeks after sowing followed by a gradual decrease. Chatterjee and Ghosh (1989) reported that in black gram, the activities of assimilatory enzymes glutamine synthetase (GS), glutamate synthase (GOGAT) and glutamate dehydrogenase (GDH) increased during germination and nodule development and maximum activity of GS/GOGAT was noticed in 5 week old nodules. It was further reported that ATP content and energy charge were maximum in 3 week old nodules.

Rao *et al.* (1989) reported that application of 20 kg N/ha increased nitrate reductase activity (NRA) in all parts of the plant and it was highest in nodules. Seed inoculation with *Rhizobium* had no effect on NRA, and increasing N rates increased NRA both in plants grown from inoculated and uninoculated seeds. Mathur *et al.* (1998) reported that NRA plays an important role in the formation of functional nodules and, rhizobial inoculation increased it in 16 test genotypes of black gram studied. Further it was reported that the genotypes which showed higher NRA also showed higher leghaemoglobin content in the nodules. Ramanujam *et al.* (1998) reported that treatment of salicylic acid increased NRA at lower concentrations but inhibited at higher concentration. Selvaraj and Kumari (1999) reported that nodule segments treated with solutions containing molybdenum, tungsten and nitrate increased NRA 2–4 fold while in nodule segments treated with ammonia solution the nitrate reductase activity decreased with increasing concentration of ammonia.

Factors Effecting Nodulation

The legume-*Rhizobium* association is a natural and dynamic biological process, and various factors especially the inoculation conditions, host factors, symbiont factors, agricultural practices and other soil microorganisms greatly influence the symbiotic association.

a) Inoculation Conditions and Field Practices

For studying the effect of various factors on nodulation, inoculations are made under controlled conditions. The inoculum concentration and other conditions of inoculation also effect nodulation. Rao and Sharma (1980) reported that inoculum concentration at the rate of 2.3×10^4 cells/seed gave highest number of nodules per plant. Rasal and Patil (1989) reported that increasing the weight of inoculum (10^7/g) per kg seed increased nodulation. Ahmed (1999) reported that seed pelleting with *Rhizobium* had more beneficial effect. Singh, A. (1984) reported that presowing chilling treatment of seeds increased nodulation. Sengupta *et al.* (1984) reported that presowing treatment of black gram seeds with water, disodium hydrogen phosphate and polyethylene glycol had increased the growth of seedlings but had no effect on nodule numbers and weight.

Various cultivational practices like type of ploughing, method of sowing, duration and amount of water supply and other factors also influence nodule formation. Pratibha *et al.* (1996) reported that, of the different types of tillage systems used for preparing the field, tractor drawn mould board plough + discing treatment enhanced nodulation of black gram cultivated in rice fallows. Kumar and Sharma (1989) studied the effect of row spacing and seed rate on nodulation, growth and yield of black gram, and reported that nodulation and growth was highest at 30 cm row spacing. The highest grain yield per plant was obtained with 20 kg/ha seed rate mainly because of significantly higher number of nodules and dry weight of root and shoot/plant. However, grain yield per hectare was lowest in the 20 kg/ha seed rate and 30 cm spacing because of low plant populations. Singh N.P. *et al.* (1987)

reported that flooding of black gram field at the seedling or flowering stage for 24, 48, 72 or 120 h markedly decreased the nodulation. The effect increased with increasing duration of flooding, and the plants at seedling stage are more susceptible than at flowering stage.

b) Host Factors

All cultivated varieties of black gram are nodulated, but host variety had significant effect on the number of nodules formed and their fresh weight. The nutritional status of the host plays an important role in nodulation and restricted supply of essential nutrients repress the formation of new nodules and causes slower maturation of already formed nodules (Jongruaysup *et al.*, 1993). The nodules are generally larger, pink in colour and form rapidly when plants are in light (Saran, 1995) and days shorter than 11 hours decrease nodulation (Srivastava and Mathur, 1980). The effect of light on nodulation is an indirect one basing on photosynthetic ability of the plant because the developing nodules require the photosynthetic assimilates. Defoliation and shading adversely affect nodulation because of their adverse effect on photosynthesis. Greater leaf area below the 6th node increase nodulation and leaf N-content, while the leaves above 7th node influence the pod yield (Rao, 1995). Bandyopadhyay (1988) reported that host variety had significant effect on the number of nodules formed and their fresh weight. The plant type also influence nodulation. Rao and Bhadur (1982) reported that nodules formed by left handed seedlings were larger in size than right handed seedlings. Ṣingh, N. *et al.* (1993, 1994) evaluated 20 black gram genotypes with *Rhizobium* inoculation and other conditions, and reported that genotypic variation was less than corresponding phenotypic variation in various physiological parameters.

Biochemical changes in the host on entry of *Rhizobium* strain are responsible for formation or nonformation of nodules by the infected strain. Rao and Iswaran (1976) reported that the biochemical changes in the legumes (including black gram) infected by *Rhizobium* vary significantly depending on the rhizobial strain, and these changes are correlated with resistance or susceptibility of the legumes to the given strain.

Siripin *et al.* (1991) screened 66 lines of black gram for nitrogen fixing ability and found that effectiveness and infectiveness varied very much in black gram lines. U Thong-2 variety which formed a large number of nodules fixed only 24.4 kg N/ha., whereas black gram line R 200003, with relatively less number of nodules per plant, fixed highest amount of nitrogen (78.8 kg N/ha.).

Induced mutations in host genotypes also result in variation in nodulation of mutant varieties. Oblisami *et al.* (1973) reported that of the 8 mutants of CO I variety of black gram, 6 mutants showed lesser number of nodules than that of the parent and, in one of the mutant (M-6) significant suppression of nodulation of lateral roots was observed indicating a change in the interrelationship between the plant root and *Rhizobium* in the soil. Singh and Raghuvanshi (1985, 1987) reported that a pentaphyllous mutant of black gram showed better nodulation than the parent plant. Mahna *et al.* (1990) reported that in dwarf mutants nodulation decreased while in other mutants it increased.

Bell *et al.* (1990) studied the growth and nodulation of 22 tropical food legumes in dilute solution culture and reported that the growth of some legumes including black gram in solution culture was relatively poor due to failure of inoculated *Bradyrhizobium* to effectively nodulate or excessive uptake of phosphorus and/or manganese by the plant. Mahna *et al.* (1991) reported that sodium azide treatment showed phytotoxicity and reduced nodulation.

c) Effect of Rhizobium Strains

Rhizobium strains isolated from the host differ in their infectiveness, effectiveness and in cultural and biochemical characteristics depending upon the region, host variety and other factors. The symbiotic efficiency clearly vary with strain differences and host variety and selection of compatible symbiont pairs is essential for proper nodulation and yield (Bandyopadhyay, 1991). Bushby (1984) studied the effect of three *Rhizobium* strains at three different sites and found that the performance of the strains varied with soil type of the site and attributed the differences to the competition by native rhizobia. Bhuiya *et al.* (1986) compared the effect of four strains and recorded that one among the four showed significantly higher positive effect with respect to shoot, root, total dry matter, and grain yield. Ahlawat *et al.* (1992) reported that the strains producing bacteriocin have a competitive edge over non-bacteriocin producing strains in nodulating the host plants. They found that when black gram and green gram were inoculated with a bacteriocin producing strain (GR4-209) and a standard normal strain (S-24), the proportion of nodules formed with bacteriocin producing strain was twice as much as those formed by standard normal strain. Khan *et al.* (1999) studied the occupancy of the nodules by the inoculated strain by immunoblot assay in different legumes, and reported that in black gram the occupancy of the nodule by inoculated strain is 73% only, and the rest is occupied by indigenous strain.

Kumar and Srivastava (1994) reported that nitrogen fixing efficiency of *Rhizobium* isolates is highly influenced by agroclimatic conditions prevailing at a particular site and local isolates usually perform better. Basha and Vivekanandan (2000) reported that *Rhizobium* strains isolated from plants growing in virgin soil were found to be more virulent and performed better than *Rhizobium* strains isolated from plants growing in garden soil, polluted soil and saline soil.

During the process of nitrogen fixation hydrogen is evolved, and the amount of hydrogen evolved is dependent upon the presence or absence of an enzyme known as uptake hydrogenase (Hup). The presence or absence of Hup enzyme is a strain specific character and the strains positive for the enzyme are described as Hup+ strains and those that lack it are called Hup– strains. Hup+ strains are generally considered as more efficient in terms of energy utilization in nitrogen fixation. Uptake hydrogenase activity of *Rhizobium* strains forming nodules on black gram and their effect was studied by a number of workers. Dadarwal *et al.* (1982) studied the effect of inoculation of Hup+ strain (S-24) and a Hup– strain on leghaemoglobin production in nodules of six hosts of cow pea miscellany group including black gram, and reported that the nodules formed by Hup– strain produced 1.5 to 2 times more leghaemoglobin than the nodules formed by hup+ strain. However, Hup– strain was less effective than hup+ strain in black gram. Pahwa and Dogra (1983) reported that plants inoculated with hup+ strain produced greater N- content and dry matter than the plants inoculated with Hup– strain. Garg *et al.* (1983) studied the effect of five strains of cow pea rhizobia on five cow pea miscellany hosts and reported that hydrogen uptake was observed only in black gram infected with four strains of *Rhizobium,* and concluded that expression of uptake hydrogenase was under the control of host plant. Dadarwal *et al.* (1985) screened the *Rhizobium* strains of summer legumes including black gram and found that 56% of the strains were Hup+. They further reported that on inoculation with Hup+ and Hup– strains under sterilized as well as unsterilized soil conditions, the host species were predominantly nodulated by Hup+ strains. Sindhu and Dadarwal (1986) reported that expression of nitrogenase under cultural conditions is a reliable method for identification of Hup+ strains.

The presence or absence of uptake hydrogenase is not an absolute character for a given strain, and Hup– strains can be obtained from Hup+ strains and vice versa. Garg and Dadarwal (1991) derived

Hup– mutants from Hup+ *Rhizobium* strain (S-24) and inoculated black gram plants with both mutant and parent strains. It was observed that acetylene reduction activity (ARA) of Hup– mutants was at par with that of the parent in nodules. Even though both Hup+ parent and Hup– mutants showed similar nitrogenase activity in nodule tissue, under explanta conditions, Hup– mutants showed very low nitrogenase activity. Garg and Dadarwal (1993) obtained Hup– strains from a Hup+ strain (S-24) and Hup+ strains from a Hup– strain (M-11), and studied the symbiotic efficiency of Hup+ and Hup– isogenic derivatives on three hosts of cow pea miscellany group including black gram. They reported that there was no relationship between relative efficiency (as determined by plant nitrogen and plant dry weight) and symbiotic efficiency (as determined by hydrogen evolution). Dudeja *et al.* (1995) studied the symbiotic efficiency of Hup+ and Hup– rhizobia in black gram and green gram under field conditions, and reported that Hup+ rhizobia (homologous or heterologous) were not superior to Hup– strains of rhizobia in increasing nodule number, nodule mass, total dry weight of the plant and grain yield. From the above, it is clear that effect of Hup activity is more related to the efficiency of energy utilization during nitrogen fixation but not on nodulation characters. It may be because, as Garg and Dadarwal (1993) opined "the energy and carbon supply to nodules is regulated by plants depending upon the requirements of nodules, and H_2 recycling as such is not contributing significantly in the energy saving system".

Rhizobium strains can be subjected to mutagens to yield mutants and the possibility of obtaining efficient mutant strains of *Rhizobium* was studied by Sudarsanam and Prasad (1972). They subjected *Rhizobium* strains to x-ray irradiation and used the mutants for inoculation of black gram and two other legumes viz. green gram and horse gram. They found that the effect of mutant strains was variable. Some mutant strains increased nitrogen content of the root and nodule while other mutants have shown reduction in their ability to fix nitrogen by varying degrees, indicating that the mutations occur randomly.

d) Effect of Pesticides

Seed dressing with fungicides and insecticides to control seedling diseases, and use of herbicides to control weeds in the field is a normal agricultural practice, and the effect of the pesticides on nodulation by *Rhizobium* in black gram has been studied by a number of workers.

The effect of fungicides on nodulation was quite variable. Ganguli *et al.* (1975) studied the effect of five fungicides on nodulation and reported that seed dressing fungicides, fytolan and agrosan GN are highly toxic to *Rhizobium* in culture and completely inhibited nodulation in excised root culture of black gram. Aureofungin, brassicol and dithane Z-78 were inhibitory only at high concentrations but showed stimulatory effect on in vitro growth of *Rhizobium* at low concentrations. These fungicides also showed inhibitory effect on nodulation when incorporated in the rooting medium. Prasad (1985, 1986) reported that cuman-L had little or no effect on nitrogenase activity and spraying of calixin on plants did not inhibit nodulation or nitrogenase activity. Longiswaran *et al.* (1991) reported that seed treatment with carbendazim did not affect nodulation, while Ghosh (1995) reported that seeds inoculated with *Rhizobium* and treated with fungicides (carbendazim and emissan) significantly decreased nodulation.

The insecticides appear to have little effect on *Rhizobium*. Swaminathan and Prasad (1976) reported that two granular pesticides viz. quinalphos and carbofuran had no effect on nodulation, growth, dry weight and total nitrogen content of black gram plants. Prasad (1985) reported that metasystox, an insecticide, had reduced nitrogenase activity by 20% in detached root nodules. Sarkar *et al.* (1988) reported that inoculation of black gram seeds with *Rhizobium* and application of insecticides significantly increased crop yield. Singh and Mathur (1989) reported that phorate, an organophosphate insecticide, increased nodulation and plant growth. Soam and Agrawal (1989) reported that black gram plants sprayed with dimethoate decreased nodule numbers per plant by 63% and also decreased leghaemoglobin concentration, nitrogensase activity and size of bacterial region in the nodule. Mathur *et al.* (1990) reported that foliar application of rogor (dimethoate) though inhibitory initially resulted in increase in growth, nitrogen concentration and nodulation.

Prabhakaran and Ramasamy (1990) reported that, of the seven seed dressing chemicals tested, bavistin, a systemic fungicide and dimethoate, an insecticide, gave better nodulation and seed yield than carbosulfan, chlorpyrofos, phosphamidin, metasystox and nuvacron, but nodulation and growth was highest when seeds are inoculated with *Rhizobium* alone.

The effect of herbicides was also studied on nodulation and survival of *Rhizobium*. Sud and Gupta (1972) reported that two herbicides viz. TCA (Trichloroacetic acid, sodium salt) and Dalapon (2, 2-dichloropropionicacid, sodium salt) showed stimulatory effect on growth of *Rhizobium* strains from black gram and eight other legumes at 0.1 mM and 1.0 mM concentrations, and at 10 mM retarded the growth but inhibitory effect was never more than 50%. Venkateswarlu *et al.* (1988) studied the effect of fluchloranil and pendimethalin, and reported that incorporation of the herbicides in soil at pre-emergence and 20 days after sowing depressed nodulation and nodule dry weight but application of the herbicides at 40 days after sowing significantly increased nodule number per plant than in untreated plots. Gupta and Pandey (1992) studied the effect of 6 herbicides *viz.* fluchloranil, pendimethalin, alachlor, dalapon, isoprothron and 2, 4-D on survival of *Rhizobium* in soil and reported that minimum inhibitory concentration (MIC) of different herbicides on *Rhizobium* varied from 7000 to 15,000 ppm, and black gram was more sensitive to herbicides than *Rhizobium*.

e) Effect of Soil Fumigation

The effect of soil fumigation with sulphur dioxide (SO_2) on growth and nodulation black gram was studied by Lalman and Singh (1989). Soil fumigation continuously for 10 days (with 2h/day) or intermittent fumigation adversely affected the growth of the plants and nodulation. It is regarded that acid precipitation in soil due to SO_2 might have reduced bacterial population which in turn affected the formation of root nodules.

Udaiyan *et al.* (1995) reported that soil fumigation with formaldehyde has little effect on nodulation in the early stages but showed adverse effect during later stages.

f) Effect of Fertilizers and Micronutrients

The effect of fertilizers, especially of nitrogen and phosphorus fertilizers, and also the effect of micronutrients, especially molybdenum and iron, on *Rhizobium* — black gram association has been studied extensively.

Kandasamy and Prasad (1978) reported that nitrate nitrogen content of rhizosphere soil has positive effect on rhizobial population in soil. Reddy *et al.* (1978) reported that application of 15 kg N/ha as basal dose along with *Rhizobium* inoculation was superior in improving growth and yield over *Rhizobium* inoculation alone. Kaur (1991) reported that *Rhizobium* inoculation increased yield and also increased the response of the plants to fertilizer application. Upadhyay *et al.* (1991) reported that application of N-fertilizers in conjunction with *Rhizobium* inoculation was the best treatment for increase of yield in black gram. Tripathi *et al.* (1994) studied the relative efficiency of *Rhizobium* inoculation and nitrogen fertilizers on growth and yield of kharif pulses including black gram and found that the effect of inoculants was similar to that of fertilizer treatments and so the cost of N fertilizer could be saved by the use of *Rhizobium* inoculation alone. Sharma *et al.* (1999, 2000) reported that application of nitrogen and inoculation with *Rhizobium* increased nodulation and nodule dry weight/plant, crop growth and relative growth rate, but no significant effect on protein content of seed and straw.

Iswaran *et al.* (1969) reported that application of phosphate had little effect on black gram, but Ravenkar *et al.* (1973) reported that the application of superphosphate increased nodulation, growth and yield of black gram. Jamwal *et al.* (1989) reported that seed inoculation with *Rhizobium* or application of 60 kg P_2O_5/ha had shown similar effect on increased nodulation, growth and yield. Sarkar and Mukherjee (1991) reported that application of phosphorus at the time of sowing increased the number and dry weight of nodules, and that applied at the time of flowering increased the number of pods per plant. Sudhakar *et al.* (1989) and Singh and Sudhakar (1991) reported positive effect of phosphate on nodulation, N-content of the plant and yield.

The positive effect of combined application of nitrogen and phosphorus has been reported by a number of workers (Chakraborty *et al.*, 1984; Thakur and Negi, 1985; Negi and Thakur, 1985; Ramaswamy *et al.*, 1985; Mathan *et al.*, 1996). Bhalu *et al.* (1995) reported that though the application of N and P fertilizers increased yield, net return was highest with seed inoculation with *Rhizobium* over the application of N and P fertilizers.

The positive effect of NPK fertilizers on nodulation, growth and yield of black gram was reported by a number of workers (Mahadkar and Saraf, 1988a, b, c; 1991a,b; Dwivedi *et al.*, 1990, 1993; Singh, A.K. *et al.*, 1993 a, b). However, Gupta (1989) reported that the application of 10 kg N, 40 kg P_2O_5 and 20 kg K_2O/ha in different combinations with or without *Rhizobium* inoculation had no significant effect on seed yield of black gram following potatoes grown with high NPK rates. Srivastava and Srivastava (1994) reported that NPK fertilizers at higher rates reduced yields due to increased vegetative growth, reduced pod formation, reduced nodulation and nitrogenase activity. However, it was reported that the application of NPK fertilizers at lower rates had positive effect on growth and yield.

Khandekar *et al.* (1985) reported that application of phosphate and sulphate fertilizers increased growth, nodulation and yield of black gram. They observed synergistic effect of phosphorus and sulphur in respect of dry weight of roots and nodules. Ghosh (1994) reported that application of sulphur in the form of gypsum had positive effect on nodulation of black gram. Thakur (1999) reported significant increase in seed and straw yield with seed inoculation and application of phosphorus and sulphur.

Jayakumar *et al.* (1997) reported that *Rhizobium* inoculation, lime pretreatment and application of coir pith compost synergistically increased the growth and nodulation significantly.

Molybdenum (Mo) is a micronutrient essential for plant growth, and also an important constituent of enzyme nitrogenase. Its effect was studied by a number of workers. Dhillon *et al.* (1977) reported that seed inoculation with *Rhizobium* cultures had no significant effect, but addition of Mo to the soil at the rate of 14 kg/ha showed positive effect on growth of the plant. However, they found that molybdenum-*Rhizobium* interaction was not significant. Kler *et al.* (1984) reported that seed treatment with sodium molybdate had significant effect on growth of black gram but *Rhizobium* inoculation showed no significant effect either alone or in combination with Mo. Though the above reports emphasized the positive effect of Mo on plant growth rather than on nodulation, some workers reported positive effect of Mo on nodulation also. Paricha and Kar (1983) reported that the seeds inoculated with *Rhizobium* and treated with Mo gave rise to plants with increased nodulation. Tiwari *et al.* (1989) also reported that treatment of Mo increased positive effect of *Rhizobium* inoculation. Daterao *et al.* (1994) reported that seed treatment with *Rhizobium* and Mo, and application of 10 kg N gave higher yield than application of 20 kg N alone. Kishor *et al.* (1991) reported that application of Mo and iron increased the number, dry weight and N-content of nodules. Maurya *et al.* (1993) reported that when *Rhizobium* inoculated seeds were sown in soils with varying concentration of Mo and Fe, the nodule weight, seed yield and nitrogen uptake increased with increased rate of iron, and Mo application increased up take of phosphorus and calcium. Jongruaysup *et al.* (1993, 1997) reported that low levels of seed Mo had no effect on nodule initiation or the number of nodules, but adversely affected bacteroid concentration, leghaemoglobin content and fresh weight of the nodules already formed. It is further reported that low Mo in seeds resulted in slower maturation of early initiated nodules, and repression of formation of new nodules. The distribution pattern of Mo significantly differed in nodulated and nonnodulated plants. Chaudhary and Das (1996) reported that application of phosphorus, sulphur and molybdenum significantly increased canopy, nodule count and yield of rainfed black gram.

g) Effect of Growth Regulators

Various growth regulators were detected in the nodules of black gram, and the effect of growth regulators on nodules is highly variable.

Kandasamy and Prasad (1975a, 1976a, b, 1977) reported that spraying of growth hormones — IAA and gibberellic acid — increased IAA content in rhizosphere soil and rhizobial population in soil, but increased rhizosphere rhizobial population showed no positive correlation with nodulation of black gram. Jaiswal *et al.* (1981) identified four cytokinins viz. zeatin, zeatin riboside, isopentyl aminopurine and its riboside, in the nodular tissue of black gram. Srivastava and Mathur (1981) reported that gibberellic acid regulated nitrate reductase activity in the nodules of black gram by enhancing its biosynthesis. Rizvi *et al.* (1984) studied the effect of four cytokinins viz. zeatin, zeatinriboside, isopentenyladenine and isopentenyl adenosine (IPAS) on nitrogenase activity in the nodules of black gram and reported that IPAS was the most effective cytokinin which stimulated nitrate uptake as swell as nitrate reductase activity. Prasad (1986) reported that spraying of black gram seedlings with cytozyme (growth regulator with chelated trace elements) had no effect on nodulation but slightly increased nitrogenase activity. Ghosh (1996) reported that different growth regulators *viz.* IAA, NAA, GA or triacotanol applied to black gram either as soil drench or foliar spray did not significantly effect nodulation. Hemalatha (1998) reported that nodulation and auxin content of black gram nodules increased due to kinetin treatment. Ramanujam *et al.* (1998) reported that the number of nodules, their leghaemoglobin content, protein content and nitrogenase activity were reduced by seed treatment with salicylic acid. Mahla *et al.* (1999a, b, c) studied the effect of spraying black gram plants with 2 ppm mixatol and/or 20 ppm NAA, together with *Rhizobium* inoculation and phosphate application. They reported that spraying of NAA had greater positive effect on nodule number and their fresh weight than mixatol and the effect was greatest with the combination. They further reported that the effect of growth hormones was enhanced by application of phosphorus.

h) Effect of Radiation

Adverse effect of UV-radiation on nodulation in black gram, and partial alleviation of its effect by *Rhizobium* inoculation and treatment with plant growth hormones was reported.

Singh, V.P. *et al.* (1993) studied the recovery of radiation induced damage by growth promoting hormone, gibberllic acid, in black gram and reported that plants raised from black gram seeds irradiated with 15 kr, 30 kr and 45 kr doses of gamma rays showed slight reduction in nodule number and dry weight along with morphological and cytological damage. When irradiated seeds were post treated with 100 ppm and 250 ppm concentration of gibberllic acid, the plants showed significant recovery from radiation induced damage and nodulation and nodule weight was even increased over control at 15 kr and 30 kr irradiated seeds.

Singh, A. (1997) reported that increased UV-B radiation reduced growth, net photosynthetic activity and nitrogen fixation in various tropical leguminous crops including black gram. In black gram, the effect of radiation was maximum on net photosynthetic activity followed by nitrogenase activity.

Daniel *et al.* (2000) reported that in seedlings of black gram plants exposed to UV-B radiation, there was significant reduction in plant growth, nodulation, leghaemoglobin content and nitrogenase activity. They further observed that seedlings inoculated with *Rhizobium* enhanced the growth and nodulation even in irradiated plants, thus partially alleviating UV-B stress effects on the plants.

i) Effect of Weeds and Weed Plant Extracts

Interaction of plants growing in an ecosystem is quite common, and the effect of weeds and some plant extracts on nodulation of black gram was studied by some workers. Bhaskar *et al.* (1984) reported that *Celosia argentea*, a common weed of leguminous crop fields, reduced significantly the nodulation of black gram. Verma *et al.* (1985) reported that spraying of black gram plants with leaf extracts of *Clerodendrum fragrans* and *Aerva sanguinolanta* and root extracts of *Boerhaavia diffusa* decreased natural infection with mung bean yellow mosaic virus and increased nodulation. Bose *et al.* (1978) reported that treatment of black gram seeds with *Moringa* leaf extract increased nodulation and nitrogenase activity. Manikandan and Hakim (1999) reported that black gram seeds treated with *Parthenium* root extracts increased seed germination and number of nodules produced per plant.

Interaction of *Rhizobium* with other Microorganisms

Rhizobium is a very common soil bacterium, and naturally it interacts with other soil microorganisms, especially VAM fungi, plant pathogens, soil microflora and nematodes. Some of the interactions are associative and help the host, while some are antagonistic and adversely affect the nodulation of the host. Further, *Rhizobium* inoculation and establishment of symbiotic relationship cause significant changes in the biochemical composition of roots and consequently the composition of root exudates both qualitatively and quantitatively, which had a great influence on rhizosphere microflora.

Rao (1976) reported that upon inoculation with *Rhizobium* there was an increase in the exudation of nonreducing sugars, orthodihydroxy phenols, amino-nitrogen, polygalacturonase and pectin methyl esterase and a decrease in reducing sugars and total phenols. The root exudates of plants inoculated with homologous *Rhizobium* differed quantitatively from those of plants inoculated with heterologous *Rhizobium* and of uninoculated plants. The changes in the composition of root exudates effect the soil microbial populations and their interaction with *Rhizobium*.

***a) Rhizobium* — VAM Fungi**

Legumes are unique in having symbiotic association with rhizobia which fix atmospheric nitrogen, and with vesicular-arbuscular mycorrhizal (VAM) fungi which helps the host in accumulation of phosphorus. The effect of VAM fungi on nodulation of legume crops is a subject of great agricultural importance, and studied in a number of crop plants including black gram.

The general effect of VAM fungi with *Rhizobium* was associative and increase nodulation. Khan *et al.* (1988) reported that root infection by indigenous VAM fungi resulted in improvement of nodulation and this positive effect was attributed to better uptake of phosphate. Rao and Rao (1993) reported that dual inoculation of black gram with *Rhizobium* and VAM fungi (*Glomus mosseae* or *G. epigaecum*) showed significant increase in growth, phosphorus and nitrogen uptake, nodulation, leaf chlorophyll, total soluble sugars, total phenols and free aminoacid content in roots compared with those inoculated with *Rhizobium* alone. Sankaranarayanan and Sundarababu (1998) reported that dual inoculation of *Rhizobium* and *Glomus mosseae* resulted in highest nodulation of plants. Inoculation of host with *Meloidogyne incognita*, a soil nematode, reduced the number of nodules per plant, but triple inoculation with *Rhizobium*, *G. mosseae* and *M. incognita* resulted in increased nodulation similar to dual inoculation of *Rhizobium* and *G. mosseae* indicating that the adverse effect of the nematode is effectively countered by dual inoculation with symbiotic microbes. Ghosh and Poi (1998) reported that nodulation, plant growth, phosphorus uptake and microbial population in the rhizosphere were highest when black gram was inoculated with *Rhizobium*, *Glomus fasciculatum* and *Bacillus polymyxa* (phosphate solubilizing bacterium). Kulshreshta and Khan (1999a) studied the impact of sulphurdioxide (SO_2) on root colonization by VAM and *Rhizobium*, and reported that root symbionts are capable of alleviating to a greater extent the adverse effects of SO_2 on plants, and it was attributed to better nutritional status of the inoculated plants. It was further reported that individually the positive effect of VAM (*Glomus caledonicum*) was more than that of the root nodule bacterium. Kulshreshta and Khan (1999b) studied the effect of fly ash from coal based power plants on nodulation and root colonization by VAM fungi in black gram. They reported that fly ash at 20%, 40% and 60% levels increased plant growth, and at 80% and 100% levels reduced the plant growth significantly. The positive effect of inoculation with *Rhizobium* and VAM (*Glomus caledonicum*) was more prominent on plants growing on low levels of fly ash, and at higher levels of fly ash, the symbionts reduced the adverse effects due to fly ash. Kulshrestha and Khan (2002) studied the effect of ozone on root nodulation and VAM colonization, and reported that ozone treatment completely inhibited nodulation in uninoculated plants while nodulation occurred in *Rhizobium* inoculated plants. The nodule number increased with dual inoculation with *Rhizobium* and VAM fungus. These studies clearly brought out the synergistic effect of *Rhizobium* and VAM in increasing the growth of the host plant, and in alleviating the adverse effects of other factors like SO_2, ozone and nematode infection.

b) Effect of Rhizosphere Fungi

Jain and Gupta (2002) studied the effect of 16 fungi isolated from rhizosphere of black gram on nodulation of the host and reported significant reduction in nodule number when soil was mixed with the homogenized fungal mats of *Mucor lutens, Cephalosporium coremioides, Trichoderma lignorum, Aspergillus sydowi, A. flavus, A. terreus, A. luchuensis, A. niger, Penicillium citrinum, Paecilomyces fusisporus* and *Fusarium roseum*. The maximum reduction was recorded in case of *A. niger*. The fungi that showed least effect on nodulation include *Myrothecium roridum, Cladosporium herbarum* and *Curvularia lunata*.

c) Interaction with Other Bacteria

Studies on interaction between *Rhizobium* and other bacteria are comparatively very few. Jain and Rewari (1974) isolated some bacteria and actinomycetes from the seeds of black gram and found them to show antagonism to *Rhizobium* sp. Kandasamy and Prasad (1975b) reported that in the rhizosphere soil of black gram other legumes, sprayed with growth regulators, the population of *Streptomyces* species antagonistic to *Rhizobium* decreased significantly when compared to unsprayed controls. They indicated the possible scope of foliar application of growth regulators as a means of controlling the occurrence of antirhizobial actinomycetes in the rhizosphere. Ranagarajan *et al.* (1984) isolated a lysogenic strain of *Streptomyces* sp. (NSA 4) from the nodule surface of black gram and reported that it effectively inhibited the growth of *Rhizobium* in nutrient agar and broth cultures.

d) Interaction with Plant Pathogens

A number of viruses and fungi infect black gram, and the interaction of *Rhizobium* and plant pathogens has also been studied by a number of workers.

i) Viral Pathogens

Singh and Mall (1974) studied the effect of two strains of Arhar Mosaic virus (ASMV and AMMV) on growth, nitrogen value, nodulation and nitrogen fixation of three legume crops including black gram and found varying effects on the three hosts. The virus infection increased nodule number and their fresh weight in black gram, while nodulation decreased in other two crops *viz.* green gram and cowpea. The total nitrogen content of infected plants increased in diseased plants of black gram, while it decreased in the other two crops due to virus infection. Singh and Mall (1978) studied the effect of nutrient solution containing different concentrations of nitrogen on growth and nodulation of black gram seedlings inoculated with urd bean mosaic virus (UBMV) and reported that the growth response of infected plants was similar to that of controls indicating that the virus has no effect on nodulation, growth and nitrogen content of the plant. Singh A.K. *et al.* (1985, 1986) reported that infection of black gram with urd bean mosaic virus decreased the leghaemoglobin content of nodules, but percentage dry matter of diseased plants increased. Singh, A.K. and Srivastava (1985, 1988) studied the nodular physiology and biochemical changes in black gram plants infected with urd bean mosaic virus and reported that in the nodules of infected plants total nitrogen, nitrite and nitrate nitrogen, protein and total free aminoacid content increased and ammonical nitrogen decreased when compared to them in nodules of healthy plants. It was further reported that the higher nitrogen content of various parts of infected plants over healthy plants occurred irrespective of *Rhizobium* inoculation. Rao *et al.* (1985) reported that in all the 9 cultivars of black gram infected with *Dolichos* enation mosaic virus (DEMV) the number of nodules, their size and weight decreased appreciably, but nitrogen content increased.

ii) Fungal Pathogens

Konde *et al.* (1980) reported that under cultural conditions *Rhizobium* strains of black gram showed antagonistic action to three root infecting pathogenic fungi *viz. Macrophomina phaseoli, Sclerotium rolfsii* and *Fusarium oxysporum*. But on seed inoculation of the fungi individually along with *Rhizobium*, nodulation and nitrogen fixation decreased. However, the adverse effect was partial when more than one fungus was used in seed inoculation, which they have attributed to the antagonism among the fungi. Shahzad *et al.* (1991) reported that seed treatment with *Rhizobium* significantly reduced infection by *Macrophomina phaseolina*, but the effect was not uniform at different sites.

Siddiqui *et al.* (2000) reported that coinoculation of black gram seeds with *Bradyrhizobium* sp. and *Pseudomonas aeruginosa* significantly reduced infection by three root infecting fungi *viz. Macrophomina phaseolina, Fusarium solani* and *Rhizoctonia solani*, and by one nematode *viz. Meloidogyne javanica*.

Unlike the interaction between soilborne pathogens and *Rhizobium*, the aerial plant pathogens had adverse effect on nodulation also. Nawaz and Narayanasamy (1983) reported that infection of black gram by *Erysiphe polygoni*, causing powdery mildew disease, adversely affected plant growth and nodulation, and as disease intensity increased the adverse effect on nodulation became very prominent.

e) Effect of Nematodes

Nematodes are common and important constituents of soil fauna. They infect the plants and adversely affect the other soil flora, especially, bacteria. Black gram is attacked by soil nematode populations, especially by the species of *Meloidogyne* (root-knot nematode) and species of *Heterodera* (golden-cyst nematode).

Jain and Rewari (1977) reported that *Rhizobium* of black gram did not survive in soil infested with *Meloidogyne incognita* indicating that the nematode is antagonistic to *Rhizobium*. Chahal and Singh (1983) reported that different levels of *Meloidogyne incognita* populations in soil significantly decreased the number of nodules per plant, dry weight of the plant and grain yield. Chahal *et al.* (1988) reported that inoculation of soil with *Meloidogyne incognita* significantly reduced number of nodules per plant, and also leghaemoglobin and bacterial content of the nodules. Kumar and Vadivelu (1993) reported that *M. incognita* caused significant damage to nodulation even in plants inoculated with *Rhizobium*. Fazal *et al.* (1998) reported that inoculation of black gram individually with a nematode *Meloidogyne javanica*, and a fungal root pathogen *Rhizoctonia bataticola*, reduced plant growth and nodulation significantly. In concomitant inoculations, the reduction was greater than the additive effect of the pathogens acting independently indicating a synergistic effect on plant growth and nodulation. Tiwari (1998) reported that nodulation in black gram was adversely affected in plants infested with another nematode, *Heterodera cajani*.

Effect of *Rhizobium* Inoculation on Growth and Yield of Black Gram

The positive effect of *Rhizobium* inoculation on nodule formation, growth and yield of black gram was reported by a number of workers. Rewari (1984, 1985) reported that in the field trials conducted at different locations under "All India Coordinated Research Project" (AICRP) on improvement of pulses, the effect of *Rhizobium* inoculation on yield varied widely and ranged from 0 to 29%. Vaishya and Gajendragadkar (1982) reported that yield increase in different cultivars due to *Rhizobium* inoculation ranged from 8.44% to 42.89% over control with a mean of 21.81%. Hameed *et al.* (1988) reported that bacterization of soaked seed increased seed yield up to 33% over control. Increase in yield due to *Rhizobium* inoculation was reported as 21.5% by Khune *et al.* (1989) and 25% by Sarkar *et al.* (1989). Nagre (1982) reported that increase in yield due to *Rhizobium* inoculation is similar to the increase due to application of 10 or 20 kg N/ha. Kerni and Gupta (1991) reported that dry weight of the plant in inoculated plants was slightly higher than in plants given 15 kg N/ha. Jamwal *et al.* (1989) reported that seed inoculation with *Rhizobium* increased yield comparable to application of 60 kg P_2O_5/ha. Bhalu *et al.* (1995) reported that net return was highest with seed inoculation with *Rhizobium* over application of nitrogen or phosphorus fertilizers. The effect of *Rhizobium* inoculation on yield varied with site also (Iyer and Vats, 1985; Kaur, 1991). Ramaswamy and Oblisami (1984) reported that positive effect on grain yield and harvest index due to *Rhizobium* inoculation was significantly higher in short duration varieties. Bandyopadhyay (1991) reported that serologically distinct strains also increased yield appreciably under field and laboratory conditions. Basha and Vivekanandan (2000) reported that in virgin soil, in *Rhizobium* treated plants of *Vigna mungo* the number of pods and total dry weight of seeds increased by 58% and 57% respectively.

Ramaswamy and Nair (1965) reported that inoculation with *Rhizobium* did not increase the yield significantly and attributed it to the presence of ineffective native strain in the soil. Raza (1978) reported that positive effect of nodulation was seen mainly on growth but not on yield. Jain *et al.* (1988) compared the response of 4 legume crops *viz.* black gram, soybean, groundnut and red gram, to *Rhizobium* inoculation and reported that the yield response to *Rhizobium* inoculation was highest in groundnut followed by red gram and least in black gram.

Effect on Soil Fertility and Yield of other Crops

The positive effect of legume crops on increasing soil fertility and on other non-leguminous crops is known since long time. The extent of benefit can be judged by the yield of the subsequent crop (Tilak, 1991). The effect of black gram–*Rhizobium* association on this aspect has also been studied. Sahu (1973) reported that nitrogen status of the soil was increased appreciably when two legume crops viz. black gram and horse gram were cultivated, and the effect was more with black gram than horse gram cultivation. Sharma *et al.* (1986) studied the residual effect of leguminous crops, including black gram, on some chemical and microbiological properties of soil. It was reported that total available nitrogen content of soil increased after rabi and kharif season legume crops when compared to that in soil under cereal crops. Total phosphorus content of soil was not much different under legume or cereal crops but soils under legumes showed more available phosphorus. The soils under legumes harboured higher microbial load except *Azotobacter*. Sarkar *et al.* (1989) reported that soil nitrogen content increased by 20 kg/ha under black gram plants inoculated with *Rhizobium* when compared to the soil under uninoculated plants. Qasim *et al.* (1994) reported that cultivation of 5 legume crops, including black gram, increased soil organic matter, decreased available phosphorus, increased infiltration rate and decreased soil pH.

Positive effect of crop rotation with black gram and non-legumes was also reported. Subba Rao and Tilak (1977) reported 2–4% increase in the yield of Kalyanasona variety of wheat in fields earlier sown with black gram inoculated with *Rhizobium*, when compared to the yield of the same variety in fields earlier sown with uninoculated black gram. Positive effect on yield of wheat due to black gram-wheat rotation was also reported by a number of other workers (Singh, S.B. and Verma, 1985; Singh, V.P.N. and Uttam, 1994; Yadav, 1990). Zaman and Choudhuri (1999) reported that the yield of non-legume crops like wheat, sarson, safflower and others increased appreciably when sown in fields in which black gram was cultivated previously. Mahdkar and Saraf (1988a, b) reported that black gram crop has a pronounced residual effect on succeeding forage sorghum yield. Similar results were also reported by Dwivedi *et al.* (1990, 1993).

However, adverse effect of intercropping on black gram was reported, especially, with rice and sugarcane. Yadav (1981) reported that inter cropping of black gram with sugarcane reduced cane yield and also decreased nodulation in black gram. Mandal *et al.* (1991) reported that intercropping of black gram with rice adversely affected both nodule number and nodule dry weight per plant. Mandal *et al.* (1999) reported that growth of pulses including black gram was best as sole crop than inter cropping with rice, and it is attributed to the interception of solar radiation by rice. However, Jana *et al.* (1996) reported that the number and dry weight of nodules per plant in black gram and two other legumes (green gram and groundnut) were not significantly affected by intercropping of these legumes with rice. Patra and Poi (1988) reported that inter cropping of pulses including black gram with cereals like maize and wheat reduced nodulation and nitrogen fixation by pulses. It was reported that some of the fixed nitrogen may act as fertilizer for cereal crops increasing their vegetative growth and shading, thus further reducing legume nodulation and nitrogen fixation.

There are some reports of adverse effect on nodulation even with intercropping of different legumes. Mishra and Majumdar (1990) reported that pigeon pea intercropped with black gram or groundnut or non-legumes like rice or sesame had relatively less number of nodules per plant when compared to the pure stands.

Cross Inoculation Studies

Rhizobium isolate of black gram is placed in the 'cow pea miscellany group' and cross inoculation studies with the isolate had been carried out by a number of workers with legumes of cow pea group. Cross inoculations with *Rhizobium* isolates from other plants on black gram were also reported .

Raju (1936) carried out extensive cross inoculation studies and reported that *Rhizobium* isolates from black gram, cluster bean, cow pea, dhaincha, gram, sun hemp and wild indigo belong to cow pea bacterial plant group. Bhide (1956) reported that *Rhizobium* isolates from *Vigna catjang, Arachis hypogaea, Crotalaria juncea, Phaseolus aureus* (=*Vigna radiata*), *Sesbania aegyptiaca* nodulated black gram, and the *Rhizobium* isolate from black gram nodulated *Cyamopsis tetragonaloba, Dolichos biflorus, Phaseolus aconitifolius, Cajanus cajan, Dolichos lablab, Soja max* and *Desmodium diffusum*, but failed to nodulate lima bean (*Phaseolus lunatus*) and *Cicer arietinum*. Dadarwal *et al.* (1977) reported that *Rhizobium* isolates from green gram, groundnut, red gram and black gram showed high symbiotic promiscuity with one another in that order. Dadarwal (1980) reported that black gram was effectively nodulated by *Rhizobium* strains from *Vigna unguiculata, V. radiata, Arachis hypogaea, Cajanus cajan, Aeschynomene indica* and *Dolichos lablab,* but not by *Rhizobium* isolates from *Acacia senegal, A. tortilis, Prosopis cineraria, Dalbergia sissoo, Cicer arietinum* and *Alysicarpus vaginalis.* In reciprocal cross inoculations, the *Rhizobium* isolate from black gram nodulated *Arachis hypogaea, Cajanus cajan, Vigna radiata, Aeschynomene indica, Acacia senegal* and *Prosopis cineraria,* but not *Acacia tortilis, Dolichos lablab, Dalbergia sissoo, Cicer arietinum* and *Alysicarpus vaginalis.* Outside the cow pea group hosts, some of the black gram isolates nodulated bean and lucerne, but none of the black gram isolates could nodulate pea berseem and lentil.

Gaur and Sen (1979) carried out extensive cross inoculation studies on group specificity in *Cicer-Rhizobium* symbiosis. They tested 287 isolates of *Rhizobium* from 52 legumes, including 5 isolates from black gram on 4 genotypes of *Cicer arietinum,* and reported that none of them, including those from black gram, nodulated *Cicer* genotypes.

Basak and Goyal (1980a,b) carried out extensive cross inoculation studies with rhizobia from tree legumes and two field legumes *viz.* black gram and green gram. *Rhizobium* isolates from 15 tree legumes nodulated black gram while three of them failed to nodulate green gram. Tree legume symbionts have more affinity to black gram than green gram as reflected by the formation of greater number of nodules on black gram. The tree legume species from which the *Rhizobium* isolates effectively nodulated black gram include *Acacia farnesiana, A. nubica, A. seyal* var *seyal, Albizzia lebbek* and *Dichrostachys nutans.* In terms of grain yield the *Rhizobium* isolates from *Acacia nubica, Albizzia lebbek* and *Dichrostachys nutans* gave significantly higher yields even over the black gram isolate. When *Rhizobium* isolate from black gram was inoculated to six tree species, it effectively nodulated *Acacia nubica, Albizzia lebbek* and *Dichrostachys nutans* but failed to nodulate *Acacia farnesiana, Acacia seyal* var *seyal* and *Prosopis cineraria.*

Subba Rao *et al.* (1980) reported that *Rhizobium* isolate from *Aeschynomene aspera* form nodules only on homologous host but not on other legumes including black gram, and *Rhizobium* isolates from black gram and other cow pea group of plants cannot nodulate *Aeschynomene aspera.*

Bandyopadhyay (1988) inoculated seeds of two varieties of black gram with 10 different strains of cow pea *Rhizobium* and found that all were able to nodulate black gram but both host variety and bacterial strain had significant effect on variation in nodule number and nodule fresh weight.

Garg *et al.* (1993) reported that a *Rhizobium* strain from cluster bean (*Cyamopsis tetragonaloba*) nodulated black gram, green gram and cow pea plants both in sterilized and unsterilized soil, and acetylene reduction activity (nitrogenase activity) of the nodules formed by cluster bean isolate was equal to that of nodules formed by homologous strains.

Haffez *et al.* (1995) reported that fast growing rhizobia from *Macroptilium atropurpureum* cv siratro can nodulate black gram.

Tangcham *et al.* (1996) reported that in Thailand, collection and selection of tropical rhizobia was successful for soybean and less successful for black gram.

Conclusion

From the foregoing, it is clear that a lot of research work has been done so far on black gram — *Rhizobium* association, but still a lot has to be done on this aspect. One major aspect that requires immediate attention is the characterization of the microbial symbiont. Further studies on cultural, physiological and biochemical aspects of *Rhizobium* sp. (black gram) are needed since relatively little information is available on these aspects. Extensive studies are also needed on *Rhizobium* – soil microflora interactions, as this important aspect received very little attention so far. The cross inoculation studies, though carried out widely, should be continued for identification of better strains of *Rhizobium* for black gram nodulation, and the search for it may include tree legumes and wild legumes.

REFERENCES

Ahlawat, O. P., Sharma, H. R. and Dadarwal, K. R. 1991. Host competitiveness among *Rhizobium* strains of cowpea miscellany. *Indian Journal of Microbiology*, **31**: 147–152.

Ahlawat, O.P., Sharma, H.R. and Dadarwal, K.R. 1992. Bacteriocin producing mutants of cowpea miscellany rhizobia — role in strain competition. *Indian Journal of Microbiology*, **32**: 263–268.

Ahmed, A.S. 1999. Effect of seed pelleting on field performance of black gram. *Legume Research*, **22**: 109–112.

Allen, O.N. and Allen, E.K. 1981. *The Leguminosae — A source book of characters, uses and nodulation*, McMillan Publishers Ltd., London, UK, 812 pp.

Bandyopadhyay, S.K. 1988. Variation in host and bacteria in the nodulation of *Phaseolus* species. *Environment and Ecology*, **6**: 542–546.

Bandyopadhyay, S.K. 1991. Symbiotic efficiency — a criterion for selection of agronomically superior strains in *Vigna mungo* (L) Wilczek. *Experimental Genetics*, **7**: 26–30.

Basak, M.K. and Goyal, S.K. 1980a. Studies on tree legumes–III characterization of the symbionts and direct and reciprocal cross inoculation studies with tree legumes and cultivated legumes. *Plant and Soil*, **56**: 39–51.

Basak, M.K. and Goyal, S.K. 1980b. Studies on the biology of tree legumes *Rhizobium* symbiosis: nodulation pattern and cross inoculation trials with the tree legumes and cultivated legumes. *Annals of Arid Zone*, **19**: 427–431.

Basha, M.G. and Vivekanandan, M. 2000. Potency of rhizobial strains from different environments to increase economic productivity in some legumes. *Philippine Journal of Science*, **129**: 131–134.

Bell, R.W., Edwards, D.G. and Asher, C.J. 1990. Growth and nodulation of tropical food legumes in dilute solution culture. *Plant and soil*, **122**: 249–258.

Bhalu, V.B., Sadaria, S.G., Kaneria, B.B. and Khanpara, V.D. 1995. Effect of nitrogen, phosphorus and *Rhizobium* inoculation on yield and quality, N and P uptake and economics of black gram (*Phaseolus mungo*). *Indian Journal of Agronomy*, **40**: 316–318.

Bhaskar, A., Vyas, K.G. and Pandya, S.M. 1984. Nitrogen relationships of *Vigna mungo* in pure stands and in *Celosia argentea* mixtures. *Annals of Botany*, **54**: 799–805.

Bhide, V.P. 1956. Cross inoculation studies with some rhizobia of the cow pea group. *Indian Phytopathology*, **9**: 198–201.

Bhuiya, Z.H., Islam, M.R., Uddin, M.J. and Hoque, M.S. 1986. Performance of some *Rhizobium* inoculants on black gram (*Vigna mungo*). *Bangladesh Journal of Agriculture*, **11**: 55–63.

Bose, B., Srivastava, R.C. and Mathur, S.N. 1987. Nodulation and nitrogenase activity in *Vigna mungo* in response to seed soaking and root dressing treatments of *Moringa* leaf extract. *Indian Journal of Plant Physiology*, **30**: 362–367.

Bushby, H.V.A. 1984. Colonization of rhizosphere and nodulation of two *Vigna* species by rhizobia inoculated on to seed: influence of soil. *Soil Biology and Biochemistry*, **16**: 635–641.

Chahal, P.P.K. and Singh, I.J. 1983. Interaction between *Rhizobium* and different population levels of *Meloidogyne incognita* on black gram. *Journal of Research*, Punjab Agric. Univ., **20**: 224–225.

Chahal, V.P.S., Chahal, P.P.K. and Kapoor, S. 1988. Effect of *Meloidogyne incognita* on symbiosis between *Rhizobium* and mash (*Vigna mungo*). *Crop Improvement*, **15**: 95–97.

Chakrabarty, S.K. and Agrawal, R.L. 1990. Identification of black gram varieties IV. Utilization of *Rhizobium* strain in varietal identification. *Seed Research*, **18**: 47–53.

Chakraborty, R.N., Sengupta, K. and Bhattacharyya, K.K. 1984. Effect of foliar application of fertilizers on growth and yield of black gram. *Fertilizer Technology*, **21**: 105–107.

Chatterjee, S., Choudhuri, M.M. and Ghosh, B. 1983. Changes in polyamino contents during root and nodule growth of *Phaseolus mungo*. *Phytochemistry*, **22**: 1553–1555.

Chatterjee, S. and Ghosh, B. 1989. Ammonia assimilatory enzymes and adenylate nucleotides during germination and root nodule development of *Vigna mungo*. *Plant Physiology and Biochemistry*, New Delhi, **16**: 150–162.

Chaudhuary, P. and Das, S.K. 1996. Effect of P, S and Mo application on yield of rainfed black gram and their residual effect on safflower and soil and water conservation in an eroded soil. *Journal of the Indian Society of Soil Science,* **44**: 741–745.

Dadarwal, K.R. 1980. Host bacterium factors involved in legume symbioses. *Indian Journal of Microbiology,* **20**: 245–252.

Dadarwal, K.R., Grover, R. and Tauro, P. 1982. Uptake hydrogenase in *Rhizobium* and nodule leghaemoglobin in cowpea miscellany hosts. *Archives of Microbiology,* **133**: 303–306.

Dadarwal, K.R., Prabha, S., Tauro, P. and Subba Rao, N.S. 1977. Serology and host range infectivity of cowpea group rhizobia. *Indian Journal of Experimental Biology,* **15**: 462–465.

Dadarwal, K.R. and Sen, A.N. 1973. Inhibitory effect of seed diffusates of some legumes on rhizobia and other bacteria. *Indian Journal of Agricultural Sciences,* **43**: 82–87.

Dadarwal, K.R., Sindhu, S.S. and Batra, R. 1985. Ecology of Hup$^+$ *Rhizobium* strains of cowpea miscellany: native frequency and competence. *Archives of Microbiology,* **141**: 255–259.

Daniel, R.R., Santhaguru, K. and Gunasekaran, P. 2001. Effect of supplementary UV-B radiation on growth, nodulation and nitrogen fixation in *Vigna mungo* L. and *Vigna radiata* (L) Wilczek. *Indian Journal of Microbiology,* **41**: 157–161.

Daterao, S.H. , Babu, B.V., Deshpande, V.L. and Sattiwale, T.R. 1994. Influence of *Rhizobium* and *Rhizobium* plus molybdenum on nitrogen use efficiency and economy of legumes. *Journal of Soils and Crops,* **4**: 145–147.

De Condolle, A. 1886. "*Origin of cultivated plants*". Hafner Publications Co., New York (reprint 1967), 468 pp.

Dhillon, G.S., Kler, D.S., Walia, A.S. Chahal, V.P.S. and Gupta, R.P. 1977. Effect of molybdenum in relation to *Rhizobium* culture on the yield of some pulse crops. *Indian Journal of Agronomy,* **22**: 170–171.

Dudeja, S.S., Khurana, A.L., Sharma, P.K., Dogra, R.C. and Garg, F.C. 1995. Symbiotic efficiency of hup$^+$ and Hup$^-$ rhizobia in mung bean and urd bean under field conditions. *Indian Journal of Microbiology,* **35**: 189–194.

Duke, A. 1981. Hand book of legumes of world Economic importance. Plenum Press, New York and London, pp. 345.

Dwivedi, M., Upadhyay, R.M. and Dwivedi, G.K. 1990. Effect of inorganic, organic and biofertilizers on yield, protein and aminoacid contents of black gram and wheat grown in sequence. *Annals of Agricultural Research,* **11**: 191–198.

Dwivedi, M., Upadhyay, R.M. and Dwivedi, G.K. 1993. Effect of inorganic, organic and biofertilizers on yield and nutritional quality of black gram wheat grown in sequence. *Indian Journal of Agricultural Chemistry,* **26**: 111–122.

Fazal, M., Bhat, M.Y., Imran, M. and Siddiqui, Z.A. 1998. A disease of black gram involving *Meloidogyne javanica* and *Rhizoctonia bataticola. Nematologia Mediterranea,* **26**: 63–66.

Fred, E.B., Baldwin, I.L. and McCoy, E. 1932. "*Root nodule bacteria and leguminous plants*". University of Wisconsin Press, Madison, Wis., USA.

Gandhi, R.C. and Vyas, S.R. 1969. Studies on rhizobia from different legumes grown in Punjab, Haryana and Himachal Pradesh. *Punjab University Journal of Research,* **6**: 678–686.

Ganguli, S., Siddiqui, K.A.I. and Banerjee, A.K. 1975. Effect of anti-fungal agents on *in vitro* growth of *Rhizobium* and on the formation of nodules in excised root cultures of *Phaseolus mungo*. *Indian Journal of Experimental Biology*, **13**: 588–590.

Garg, F.C., Garg, R.P., Kukreja, K., Sindhu, S.S. and Tauro, P. 1985. Host dependent expression of uptake hydrogenase in cow pea rhizobia. *Journal of General Microbiology*, **131**: 93–96.

Garg, F.C. and Tauro, P. 1985. Seasonal and diurnal profiles of nitrogenase activity of some legumes in semiarid tropics. *Zbl. Mikrobiol.*, **140**: 199–203.

Garg, R.P. and Dadarwal, K.R. 1991. Ex planta nitrogenase expression associated with Hup character in cow pea miscellany *Rhizobium*. *Indian Journal of Experimental Biology*, **29**: 371–374.

Garg, R.P. and Dadarwal, K.R. 1993. Symbiotic effectivity of Hup^+ and Hup^- isogenic derivatives of cow pea miscellany strains. *Indian Journal of Microbiology*, **33**: 35–41.

Garg, R.P., Kukreja, K. and Garg, F.C. 1993. Efficiency of a cross infective *Rhizobium* strain in heterologous hosts. *Environment and Ecology*, **11**: 256–259.

Gaur, Y.D. and Sen, A.N. 1979. Cross inoculation group specificity in *Cicer- Rhizobium* symbioses. *New Phytologist*, **83**: 745–754.

Ghosh, G. 1996. Effect of growth regulators on nodulation in some important legumes crops. *Environment and Ecology*, **14**: 422–424.

Ghosh, G. and Poi, S.C. 1998. Response of *Rhizobium*, phosphate solubilizing bacteria and mycorrhizal organisms on some legume crops. *Environment and Ecology*, **16**: 607–610.

Ghosh, S. 1995. Effect of seed dressing fungicides on nodulation of black gram (*Vigna mungo*), green gram (*Vigna radiata*) and groundnut (*Arachis hypogaea*) *in vivo*. *Environment and Ecology*, **13**: 369–371.

Ghosh, T.K. 1994. Nodulation and protein content of black gram as influenced by different levels and methods of sulfur application. *Environment and Ecology*, **12**: 916–917.

Gupta, A. 1989. Response of urd bean grown after potato to fertilizer application. *Indian Journal of Pulses Research*, **2**: 177–178.

Gupta, B.R. and Pandey, A. 1992. Effect of herbicides on survival of *Rhizobium* sp. in soil and germination of urad (black gram) seeds. *Journal of the Indian Society of Soil Science*, **40**: 569–571.

Gupta, K.G. 1969. Salt tolerance of *Rhizobium* sp. from some common cultivated legumes in liquid media. *Research Bulletin, Punjab Agricultural University*, **20**: 19–21.

Gupta, K.G., Agarwal, J.C. and Makkar, N.S. 1974. Sensitivity of rhizobia to vanillin. *Folia Microbiol.*, **19**: 317–321.

Gupta, K.G. and Sen, A. 1963. Utilization of combined nitrogen by *Rhizobium* sp. from common cultivated legumes in relation to their efficiencies. *Indian Journal of Agricultural Sciences*, **33**: 240–243.

Haffez, F.Y., Asad, S., Ahmad, T. and Malik, K.A. 1995. Host specificity and characterization of fast growing rhizobia from *Macropotilium atropurpureum* cv Siratro in Pakistan. *Soil Biology and Biochemistry*, **27**: 729–733.

Hameed, M.S., Murugesan, R. and Raj, S.A. 1988. Bacterization of soaked seeds of black gram with *Rhizobium* sp. *Madras Agricultural Journal*, **75**: 281–282.

Hellriegel, H. and Wilfarth, H. 1888. Untersuchungen uber die Stickstoffnahrung der Gramineen und Leguminosen. Addendum, Z. Ver. RubenZucker-Ind. Dtsch. Reichs, 234 pp.

Hemalatha, S. 1998. Effect of kinetin treatment and *Rhizobium* on auxin production in root nodules of black gram [*Vigna mungo* (L) Hepper]. *Advances in Plant sciences*, **11**: 57–61.

Hernandez, B. S. and Focht, D.D. 1984. Invalidity of the concept of slow growth and alkali production in cowpea rhizobia. *Applied and Environmental Microbiology*, **48**: 206–210.

Herrera-Cervera, J.A., Caballero-Mellado, J., Laguerre, G., Tichy, H.V., Requena, N., Amarger, N., Martinez-Romero, E., Olivares, J. and San Juan, J. 1999. At least five rhizobial species nodulate *Phaseolus vulgaris* in a Spanish soil. *FEMS-Microbiology-Ecology*, **30**: 87–97.

Iswaran, V., Kavimandan, S.K. and Kamath, M.B. 1969. Growth of phosphate treated legumes under inoculated and uninoculated conditions. *Food Fmg. Agric.*, **1**: 55–57.

Iyer, V.N. and Vats, M.R. 1985. A study on the efficiency of *Rhizobium* culture in nitrogen management of legume cereal rotation. *Seeds and Farms*, **11**: 21–26.

Jain, M.K. and Rewari, R.B. 1974. Isolation of seedborne microflora from leguminous crops and their antagonistic effects on *Rhizobium*. *Current Science*, **43**: 151–152.

Jain, M.K. and Rewari, R.B. 1976. Studies on seed coat toxicity to rhizobia of urid (*Phaseolus mungo*), mung (*Phaseolus aureus*) and soybean (*Glycine max*). *Zbl. Bakt. Abt. II*, **131**: 163–169.

Jain, M.K. and Rewari, R.B. 1977. A report on the infestation of urd (*Phaseolus mungo*) by root knot nematodes. *Indian Journal of Mycology and Plant Pathology*, **7**: 94.

Jain, N.K., Jain, H.C. and Khandekar, U.R. 1988. Response of kharif legumes to fertilizers and *Rhizobium* inoculation. *Indian Journal of Agronomy*, **33**: 347–350.

Jain, V. and Gupta, V.K. 2002. Effect of rhizosphere fungi on nodule number, shoot and root length of *Vigna mungo*. *Indian Phytopathology*, **55**: 323–324

Jaiswal, V., Rizvi, S.J.H., Mukherjee, D. and Mathur, S.N. 1981. Identification of cytokinins present in root nodules of *Phaseolus mungo* L. *Biologia Plantarum*, **23**: 342–344.

Jaiswal, V., Rizvi, S.J.H., Mukherjee, D. and Mathur, S.N. 1982. Nitrogenase activity in root nodules of *Vigna mungo*. The role of cytokinins. *Anganwadee Botanik*, **56**: 143–148.

Jamwal, J.S., Singh, K., Singh, T.K. and Singh, J.P. 1989. Effect of phosphorus and biofertilizers on yield and quality of black gram [*Vigna mungo* (L) Hepper]. *Annals of Agricultural Research*, **10**: 249–252.

Jana, T.K., Patra, A.P. and Roy, B. 1996. Effect of rice legume inter-cropping on the growth of nodules of legumes. *Environment and Ecology*, **14**: 434–438

Jayakumar, M., Eyini, M. and Velmurugan, R. 1997. Effect of seed *Rhizobium* pelleting treatment and addition of coir-pith compost to soil on growth and nodulation of fluoride-water irrigated black gram (*Phaseolus mungo*). *Indian Journal of Agricultural Sciences*, **67**: 601–603.

Jayapragasam, M. 1989. Lectin content of green gram and black gram genotypes and its relation to nitrogenase activity. *Indian Journal of Plant Physiology*, **32**: 353–355.

Jongruaysup, S., Dell, B., Bell, R.W., O'Hara, G.W. and Bradley, J.S. 1997. Effect of molybdenum and inorganic nitrogen on molybdenum redistribution in black gram [*Vigna mungo* (L) Hepper] with particular reference to seed fill. *Annals of Botany*, **79**: 67–74.

Jongruaysup, S., O'Hara, G.W., Dell, R.W. and Barrow, N.J. 1993. Effects of low molybdenum seed on nodule initiation, development and N_2 fixation in black gram (*Vigna mungo* L.). *Plant and soil*, **155-156**: 345–348.

Jordan, D.C. 1982. Transfer of *Rhizobium japonicum* Buchanan 1880 to *Bradyrhizobium* gen. nov. a genus of slow growing root nodule bacteria from leguminous plants. *International Journal of Systematic Bacteriology*, **32**: 136–139.

Jordan, D.C. 1984. The genus *Bradyrhizobium*. In *"Bergey's Manual of Systematic Bacteriology Vol.-I* (Eds. Kreig, N.R. and Holt, J.G.) Published by Williams and Wilkins Co., Baltimore, USA, pp 242–244.

Juwarkar, A. and Rewari, R.B. 1988. Synergistic effect of relative humidity and temperature on the survival of rhizobia in inoculant carrier. *Journal of Applied Bacteriology*, **64**: 465–469.

Kachroo, P. 1970. *"Pulses of India"*. I.C.A.R., New Delhi, pp 334.

Kandasamy, D. and Prasad, N.N. 1975a. Studies on certain factors influencing legume rhizobia interrelationships: Indole acetic acid level in rhizosphere soil. *Lab. Dev. J. Sci. Technol.*, **13**: 194–198.

Kandasamy, D. and Prasad, N.N. 1975b. Studies on certain factors influencing legume rhizobia interrelationships III. Antagonistic streptomycetes in rhizosphere region. *Hindustan Antibiotics Bulletin*, **17**: 115–117.

Kandasamy, D. and Prasad, N.N. 1976a. Studies on certain factors influencing legume-rhizobia interrelationships. IV. Exudation of aminoacids and sugars. *Indian Journal of Microbiology*, **16**: 88–91.

Kandasamy, D. and Prasad, N.N. 1976b. Foliar spray of growth factors on nodulation in three plants and its relation to *Rhizobium* population. *Indian Journal of Experimental Biology*, **14**: 737–739.

Kandasamy, D. and Prasad, N.N. 1977. Studies on certain factors influencing legume rhizobia interrelationships. II. Total microbial population in the rhizosphere region. *Madras Agricultural Journal*, **64**: 458–462.

Kandasamy, D. and Prasad, N.N. 1978. Studies on certain factors influencing legume-rhizobia interrelationships. V. Nitrate nitrogen in rhizosphere soil. *Mysore Journal of Agricultural Sciences*, **12**: 106–109.

Kaur, R. 1991. Role of *Rhizobium* culture in nitrogen management in legume-cereal rotation. *Annals of Agricultural Research*, **12**: 388–393.

Kay, D.E. 1979. *Crop and product digest No. 3 Food Legumes.* London, Tropical Products Institute, pp. 435.

Kerni, P.N. and Gupta, A. 1991. Effect of seed inoculation with endemic rhizobial strain on the growth and yield parameters of urd bean (*Vigna mungo*). *Research and Development Reporter*, **8**: 176–179.

Khan, A.H., Islam, A., Islam, R., Begum, S. and Huq, S.M.I. 1988. Effect of indigenous VA mycorrhizal fungi on nodulation, growth and nutrition of lentil (*Lens culinaris* L.) and black gram (*Vigna mungo* L.). *Journal of Plant Physiology*, **133**: 84–88.

Khan, M.S., Zaidi, A. and Lakhchaura, B.D. 1999. Nodule occupancy determination and *Rhizobium* strain quantification by immunoblot assay. *Indian Journal of Experimental Biology*, **37**: 813–817.

Khandekar, U.R., Shinde, D.A., Kanekar, V.S., Jambley, N.R. and Jain, N.K. 1985. Growth, nodulation and yield of rainfed black gram [*Vigna mungo* (L) Hepper] as influenced by phosphate and sulphate fertilization in vertisol. *Indian Journal of Plant Physiology*, **28**: 318–322.

Khune, N.N., Kolte, S.O., Chitriv, A.J. and Wangidar, P.D. 1989. Biofertilizers in Vidarbha – present and perspectives. *PKV Research Journal*, **13**: 168–169

Kishor, K., Maurya, B.R. and Ram, P.C. 1991. Effect of iron and molybdenum nutrition on nodulation, symbiotic N_2 fixation and grain yield of urd bean [*Vigna mungo* (L) Hepper]. *Narendra Dev Journal of Agricultural Research*, **6**: 189–193.

Kler, D.S., Dhillon, G.S. and Chahal, V.P.S. 1984. Effect of seed treatment with molybdenum and *Rhizobium* culture on growth and yield of black gram (*Vigna mungo*). *Environment and Ecology*, **2**: 185–187.

Konde, B.K., Shinde, D.B. and More, B.B. 1980. Studies on root infecting fungi and *Rhizobium* of black gram (*Vigna mungo* L.). *J. Maharastra Agric. Univ.*, **5**: 222–226.

Konde, B.K., Shinde, D.B. and More, B.B. 1982. Effect of seed diffusates of black gram on the growth of *Rhizobium* spp. *Seed Research*, **10**: 129–132.

Kulshrestha, M. and Khan, M.W. 1999a. Impact of sulphur dioxide on root colonization by VAM fungi and nodulation on black gram. *Indian Phytopathology*, **52**: 182–184.

Kulshrestha, M. and Khan, M.W. 1999b. Impact of fly ash application in soil on root colonization by a VAM fungus and root nodulation by *Rhizobium*. *Indian Phytopathology*, **52**: 185–187.

Kulshrestha, M. and Khan, M.W. 2002. Effect of ozone on root colonization by VAM fungi and root nodulation on black gram. *Indian Phytopathology*, **55**: 264–268.

Kumar, A. and Sharma, B.B. 1989. Effect of row spacing and seed rate on growth, nodulation and yield of black gram (*Phaseolus mungo*). *Indian Journal of Agricultural Sciences*, **59**: 728–729.

Kumar, A. and Srivastava, M. 1994. Survey of local rhizobial isolates for their efficiency in nitrogen fixation and biomass production in *Vigna mungo* (L.) Hepper. *Indian Journal of Plant Physiology*, **37**: 264–266.

Kumar, S. and Vadivelu, S. 1993. Inter-relationships between *Meloidogyne incognita* and *Rhizobium* sp. on black gram, *Vigna mungo* (L.) Hepper. *Indian Journal of Nematology*, **23**: 75–81.

Lahiri, K., Chattopadhyay, S., Chatterjee, S. and Ghosh, B. 1999. Spermine induced changes in nitrogenase and alpha amylase during nodule senescence of *Vigna mungo* (L). *Indian Journal of Plant Physiology*, **42**: 254–257.

Lalman, and Singh, B. 1989. Effect of SO_2 fumigation on growth and nodulation of *Vigna mungo*. *Journal of Indian Botanical Society*, **68**: 89–94.

Longiswaran, G., Rajamanickam, B. and Gopalan, M. 1991. Effect of insecticidal and fungicidal seed treatments on rhizobial inoculum and nodulation of black gram. *Madras Agricultural Journal*, **78**: 108–109.

Mahadkar, U.V. and Saraf, C.S. 1988a. Effect of various inputs on fertilizer nitrogen utilization and yield of black gram and green gram and their residual effect on succeeding fodder sorghum. *Fertilizer News*, **33**: 43–46.

Mahadkar, U.V. and Saraf, C.S. 1988b. Effect of *Rhizobium* inoculation, phosphorus and lime and methods of nitrogen application on growth and yield of spring black gram [*Vigna mungo* (L) Hepper] and its residual effect on succeeding fodder Sorghum (*Sorghum vulgare* L). *Journal of Agricultural Science*, UK, **110**: 677–679.

Mahadkar, U.V. and Saraf, C.S. 1988c. Effects of *Rhizobium*, phosphorus and time and method of nitrogen application on uptake and utilization of nitrogen by black gram. *Journal of Maharashtra Agricultural Universities*, **13**: 243–245.

Mahadkar, U.V. and Saraf, C.S. 1991a. Nutritional status and grain yield of spring urd bean as influenced by *Rhizobium* inoculation, nitrogen and phosphorus application. *Journal of Maharashtra Agricultural Universities*, **16**: 321–323.

Mahadkar, U.V. and Saraf, C.S. 1991b. Nitrate reductase in relation to nodulation and nitrogen nutrition of black gram. *Journal of Maharashtra Agricultural Universities*, **16**: 395–397.

Mahla, C.P.S., Dadheech, R.C. and Kulhari, R.K. 1999a. Effect of plant growth regulators on growth and yield of black gram [*Vigna mungo* (L.) Hepper] at varying levels of phosphorus. *Annals of Agri-Bioresearch*, **4**: 207–209.

Mahla, C.P.S., Dadheech, R.C. and Kulhari, R.K. 1999b. Effect of plant growth regulators on growth and yield of black gram [*Vigna mungo* (L.) Hepper] at varying levels of Phosphorus. *Annals of Biology*, Ludhiana, **15**: 205–207.

Mahla, C.P.S., Dadheech, R.C. and Kulhari, R.K. 1999c. Effect of plant growth regulators on growth and yield of black gram [*Vigna mungo* (L.) Hepper] at varying levels of phosphorus. *Crops Research*, Hissar, **18**: 163–165.

Mahna, S.K., Garg, R. and Parvateesam, M. 1990. Nodulation studies with induced mutants of black gram (*Vigna mungo* L.). *Mutation Breeding News Letter*, **36**: 9–10.

Mahna, S.K., Garg, R. and Parvateesam, M. 1991. Phytotoxic influence of sodium azide on the growth and nodulation in *Vigna mungo* L. *Journal of Phytological Research*, **4**: 49–54.

Mandal, B.K., Dhara, M.C., Mandal, B.B., Bhunia, S.R. and Dandapat, A. 1991. Nodulation in some legumes grown as pure and intercrops. *Indian Agriculturist*, **35**: 15–19.

Mandal, B.K., Saha, S. and Jana, T.K. 1999. Growth and resource use of rice, green gram, black gram and pigeon pea in different inter- and mixed-cropping systems under rainfed upland situation. *Oryza*, **36**: 358–363.

Manikandan, R. and Hakim, S.A. 1999. Effect of IAA and *Parthenium* crude extract on black gram and groundnut. *Advances in Plant Sciences*, **12**: 345–348

Mathan, K.K., Francis, H.J. and Ramanathan, S.P. 1996. Response of black gram (*Phaseolus mungo*) to fertilization and *Rhizobium* inoculation. *Indian Journal of Agronomy*, **41**: 74–77.

Mathur, A., Mathur, S. and Mathur, S.N. 1990. Effect of rogor (dimethoate) on growth, yield and nitrogen content in *Vigna mungo* (L.) Hepper. *Indian Journal of Plant Physiology*, **33**: 97–100.

Mathur, V.L., Kant, L. and Joshi, A. 1998. Study of nodulation, leghaemoglobin content and NRA in black gram (*Vigna mungo* (L.) Hepper). *Legume Research*, **21**: 221–224.

Maurya, B.R., Kishor, K. and Ram, P.C. 1993. Effects of iron and molybdenum on urd bean. *Journal of Maharashtra Agricultural Universities*, **18**: 128

Mishra, G.C. and Majumdar, D.K. 1990. Nodulation and nitrogen fixation in legumes and non-legumes with pigeon pea. *Orissa Journal of Agricultural Research*, **3**: 166–168.

Mpepereki, S., Wollum, A.G. II. and Makonese, F. 1996. Diversity and specificity of cowpea rhizobia indigenous to Zimbabwean soils. *Plant and Soil*, **186**: 167–171.

Nagre, K.T. 1982. Effect of *Rhizobium* and nitrogen on the yield of green gram and black gram. *Agricultural Science Digest*, **2**: 27–28.

Nawaz, R.M.S. and Narayanasamy, P. 1983. Effect of powdery mildew infection on growth and yield of black gram. *Madras Agricultural Journal*, **70**: 179–181.

Negi, S.C. and Thakur, R.C. 1985. Effect of nitrogen, phosphorus and *Rhizobium* inoculation on mash yield and N and P uptake. *Himachal Journal of Agricultural Research*, **11**: 96–99.

Oblisami, G., Ramaswamy, N.M., Rangaswami, G. and Purushothaman, D. 1973. Preliminary studies on the rhizobial nodulation in induced mutants of black gram. *Madras Agricultural Journal*, **60**: 259–261.

Pahwa, K. and Dogra, R.C. 1983. Uptake hydrogenase system in urd bean (*Vigna mungo*) *Rhizobium* in relation to nitrogen fixation. *Journal of Applied Bacteriology*, **54**: 405–408.

Paricha, P.C. and Kar, M. 1983. Improving symbiotic nitrogen fixation and productivity in black gram (*Vigna mungo* L.). *Current Science*, **52**: 1143–1145

Patra, B. and Poi, S.C. 1988. Influence of intercropping on the nodulation of legumes. *Environment and Ecology*, **16**: 418–423.

Pawar, N.B. and Ghulghule, J.N. 1980. Preliminary observations on occurrence and senescence of root nodules in some pulses. *Tropical Grain Legume Bulletin*, **17/18**: 3–5.

Prabhakaran, J. and Ramasamy, K. 1990. Effect of seed dressing chemicals on the survival of *Rhizobium* sp. and growth, nodulation and grain yield of green gram and black gram. *Madras Agricultural Journal*, **77**: 285–290.

Prasad, B.N. 1985. Effect of metasystox and cuman-L on nitrogenase activities in root nodules of *Phaseolus mungo*. *Indian Journal of Plant Physiology*, **28**: 204–206.

Prasad, B.N. 1986. Effect of calixin and cytozyme on nodulation and nitrogenase activity of *Phaseolus mungo*. *International Journal of Tropical Agriculture*, **4**: 143–145.

Prasad, D.N. 1982. Studies on the ultrastructure of black gram root nodule meristem with special reference to mitochondrial activity. *Proccedings of the Indian National Science Academy – B*, **48**: 795–800.

Prasad, D.N. and Prasad, N. 1989. Subcellular morphology of the symbiotic root nodule cells of black gram (*Phaseolus mungo* L). *Legume Research*, **12**: 153–159.

Pratibha, G., Pillai, K.G., Satyanarana, V. and Hussain, M.M. 1996. Tillage systems for production of black gram (*Vigna mungo*) succeeding rice crop. *Legume Research*, **19**: 23–28.

Pueppke, S.G. and Broughton, W. 1999. *Rhizobium* sp. strain NGR 234 and *R. fredii* USDA 257 share exceptionally broad nested host ranges. *Molecular-Plant-Microbe Interactions*, **12**: 293–318.

Qasim, M., Ahmad, K., Shah, N.Z. and Razaullah 1994. Studies on the growth and nodulation in some kharif legumes and their effects on soil properties. *Sarhad Journal of Agriculture*, **10**: 231–236.

Raju, M.S. 1936. Studies on bacterial plant groups of cow pea, *Cicer* and dhaincha. I Classification. *Zbl. Bakt. Abt. II*. **94**: 249–262.

Ramanujam, M.P., Jaleel, V.A. and Kumaravelu, G. 1998. Effect of salicylic acid on nodulation, nitrogenous compounds and related enzymes of *Vigna mungo*. *Biologia Plantarum*, **41**: 307–311.

Ramaswamy, P.P. and Nair, K.S. 1965. Effect of seed inoculation with *Rhizobium* on yield and nitrogen content of pulses under field conditions. *Indian Journal of Microbiology*, **5**: 95–96.

Ramaswamy, P.P. and Oblisami, G. 1984. Influence of duration of black gram [*Vigna mungo* (L) Hepper] on harvest index among the different genotypes. *Madras Agricultural Journal*, **71**: 246–248.

Ramaswamy, P.P., Oblisami, G. and Chellamuthu, S. 1985. Influence of nitrogen, phosphorus and inoculation on nodulation and yield in black gram (*Vigna mungo* L.). *Madras Agricultural Journal*, **72**: 658–660.

Ramaswamy, P.P., Perumal, R., Nair, K.S. and Selvaraj, K.V. 1977. Shoot, root and nodule systems in certain pulses. *Current Research*, **6**: 193–195.

Rangarajan, M., Ravindran, A.D. and Hariharan, K. 1984. Occurrence of a lysogenic *Streptomyces* sp. on the nodule surface of black gram (*Vigna mungo*). *Applied and Environmental Microbiology*, **48**: 232–233.

Rangaswami, G. and Oblisami, G. 1962. Studies on some legume root nodule bacteria. *Journal of the Indian Society of Soil Science*, **10**: 175–186.

Rao, A.V. 1976. Root exudation in relation to inoculation with *Rhizobium*. *Zbl. Bakt. Abt. II*, **131**: 77–82.

Rao, A.V. and Iswaran, V. 1976. Biochemical changes in the roots of legumes upon inoculation. *Zbl. Bakt. Abt. II*, **131**: 83–90.

Rao, A.V. and Sharma, R.L. 1980. Note on effect of different inoculum levels of rhizobia on symbiosis of soybean and black gram. *Legume Research*, **3**: 55–57.

Rao, D.V.K.N., Reddy, A.T., Reddy, D.V.R. and Pillai, R.N. 1989. Effect of N-fertilization and *Rhizobium* inoculation on *in vivo* nitrate reductase activity and nitrogen content of black gram. *Journal of Maharashtra Agricultural Universities*, **14**: 332–334.

Rao, J.S.P. 1995. Effect of source sink manipulation on nodulation, dry matter production and seed yield of black gram. *Madras Agricultural Journal*, **82**: 646–648.

Rao, K., Rama, A. and Gopalan, K.R. 1985. Effect of *Dolichos* enation mosaic virus infection on nodulation in different varieties of black gram and green gram. *Journal of Research, A.P. Agricultural University*, **13**: 23–27.

Rao, M.M. and Bhadur, B. 1982. Note on the relationship between seedling handedness and root and nodule characters in green gram and black gram. *Indian Journal Agricultural Sciences*, **52**: 706–708.

Rao, V.U. and Rao, A.S. 1993. Dual inoculation of VAM and *Rhizobium* in black gram and green gram. *Legume Research*, **16**: 119–126.

Rasal, P.H. and Patil, P.L. 1989. A study on inoculum load of *Rhizobium* on some pulses. *Journal of Maharashtra Agricultural Universities*, **14**: 234–235.

Ravenkar, H.N., Badhe, N.N. and Kadwe, R.S. 1973. Effect of different levels of phosphate on growth, nodulation and nitrogen fixation by urd (*Phaseolus mungo* Roxb.) and mung (*Phaseolus aureus* L.) *College Agric. Mag. Nagpur*, **45**: 50–55.

Raza, S.H. 1978. Studies on ecological production of two species of *Phaseolus* in red soils. *Geobios*, **5**: 20–23.

Reddy, A.S. and Mallaiah, K.V. 2002. Studies on the effect of *Rhizobium* inoculation on black gram. In "Bioinoculants for sustainable Agriculture and Forestry" (Eds. Prof. S.M. Reddy *et al.*), Scientific Publishers, Jodhpur, pp. 93–98.

Reddy, P.R., Zaheda, A.A. and Rao, L.M. 1978. Influence of basal dose of nitrogen and rhizobial inoculation on yield and nitrogen uptake in black gram (*Phaseolus mungo* L). *Indian Journal of Plant Physiology*, **21**: 210–212.

Rewari, R.B. 1984. "Summerized results of Microbiology trails" — All India co-ordinated Research Project on improvement of pulses, New Delhi, India.

Rewari, R.B. 1985. "Summerized results of Microbiology trails" — All India co-ordinated Research Project on improvement of pulses, New Delhi, India.

Rizvi, S.J.H., Rizvi, V., Mukerji, D. and Mathur, S.N. 1984. Investigation on the role of nodule cytokinins in regulation of nitrate reductase activity of *Phaseolus mungo* L. *Journal of Plant Growth Regulation,* **3**: 237–244.

Sahu, S.K. 1973. Effect of *Rhizobium* inoculation and phosphate application on black gram (*Phaseolus mungo*) and horse gram. *Madras Agricultural Journal,* **60**: 989–993.

Sankaranarayanan, C. and Sundarababu, R. 1998. Effect of *Rhizobium* on the interaction of Vesicular-Arbuscular mycorrhizae and root knot nematode on black gram (*Vigna mungo*) — *Nematologia Mediterranea,* **26**: 195–198.

Saran, B. 1995. Effect of light on nodulation and leghaemoglobin in black gram. *New Agriculturist,* **6**: 27–28.

Sarkar, A. and Mukherjee, A.K. 1991. Effect of phosphorus on yield and nodulation of green gram (*Phaseolus radiatus*), black gram (*P. mungo*) and rice bean (*Vigna umbellata*). *Indian Journal of Agricultural Sciences,* **61**: 328–331.

Sarkar, B.B., Bhattacharjee, N.R. and Roy, S. 1989. *Rhizobium* culture helps increase the yields of black gram. *Indian farming,* **38**: 13–14.

Sarkar, H.K., Basu, T.K. and Pal, A.K. 1988.Interactions of *Rhizobium* and insecticides on the yield attributes of black gram (*Vigna mungo*). *Environment and Ecology,* **6**: 27–31.

Schwent, R.M., Abe, M. and Higashi, S. 1983. Cytological study of the root nodule of *Vigna mungo* (Fabaceae). *Journal of the Arizona-Nevada- Academy of Science,* **18**: 22–25.

Selvaraj, K. and Abragam, D.A. 1999. Occurrence CA-oxalate crystals in nitrogen fixing root nodules and other tissues of some tropical legumes. *Phytomorphology,* **49**: 79–87.

Selvaraj, K. and Kumari, E.V.N. 1999. Response of *in vivo* nitrate reduction in root nodules of legumes to nitrate, ammonia, molybdenum and tungsten. *Indian Journal of Plant Physiology,* **42**: 271–276.

Senaratne, R. and Ratnasinghe, D.S. 1993. Ontogenetic variation in nitrogen fixation and accumulation of nitrogen in mung bean, black gram, cow pea and groundnut. *Biology and Fertility of Soils,* **16**: 125–130.

Sengupta, K., Bhattacharyya, K.K. and Chatterjee, B.N. 1984. Effect of pre-sowing seed treatments on the growth and yield of black gram (*Vigna mungo*). *Zeitschrift-fur-acker-und-Pflanzenbau,* **153**: 334–341.

Sengupta, K., Sengupta, C. and Bhattacharyya, K.K. 1986. Seed treatment improves productivity of black gram in West Bengal. *Indian Farming,* **36**: 12–15.

Shahzad, S., Hussain, S., Dawar, S., Ghaffar, A. and Malik, B.A. 1991. Biological control of *Macrophomina phaseolina* infection on mash bean [*Vigna mungo* (L.) Hepper]. *Pakistan Journal of Botany,* **23**: 131–134.

Shanmugam, A.S. and Rangasamy, S.R.S. 1983. Studies on nitrogen fixation in black gram [*Vigna mungo* (L) Hepper] and green gram [*Vigna radiata* (L.) Wilczek]. *Proceedings of Indian National science Acadenmy — B (Biological sciences),* **49**: 367–370.

Shanmugam, A.S., Rangasamy, S.R.S. and Rathinasamy, R. 1984. Correlation and path analysis for nodule characters in black gram and green gram. *Indian Journal of Agricultural Sciences,* **54**: 712–718.

Sharma, C.P., Gupta, B.R. and Bajpai, P.D. 1986. Residual effect of leguminous crops on some chemical and microbiological properties of soil. *Journal of the Indian Society of Soil Science*, **34**: 206–208.

Sharma, S., Sharma, C.R. and Upadhyay, R.G. 1999. Response of various strains of *Rhizobium* and nitrogen on symbiotic and physiological parameters, biochemical constituents and yield of black gram [*Vigna mungo* (L.) Hepper]. *Legume Research*, **22**: 233–236.

Sharma, S., Upadhyay, R.G. and Sharma, C.R. 2000. Effect of *Rhizobium* inoculation and nitrogen on growth, dry matter accumulation and yield of black gram [*Vigna mungo* (L) Hepper]. *Legume Research*, **23**: 64–66.

Siddiqui, I.A., Haque, E., Zaki, M.J. and Ghaffar, A. 2000. Co-inoculation of *Pseudomonas aeruginosa* and *Bradyrhizobium* sp. in the control of root infecting fungi and root knot nematode on urid bean *Vigna mungo* (L.) Hepper. *Pakistan Journal of Phytopathology*, **12**: 35–41.

Sindhu, S.S. and Dadarwal, K.R. 1986. Ex planta nitrogenase induction and uptake hydrogenase in *Rhizobium* sp. (cow pea miscellany). *Soil Biology and Biochemistry*, **18**: 291–295.

Singh, A. 1984. Enhancement in growth and yield by pre-sowing chilling treatment to seeds of *Phaseolus mungo* cv. T-9. *Annals of Botany*, **53**: 59–63.

Singh, A. 1997. Increased UV-B radiation reduces N_2- fixation in tropical leguminous crops. *Environmental pollution*, **95**: 289–291.

Singh, A.K., Choudhary, R.K. and Sharma, R.P.R. 1993a. Effect of inoculation and fertilizer levels on yield, yield attributes and nutrient uptake of green gram (*Phaseolus radiatus*) and black gram (*P. mungo*). *Indian Journal of Agronomy*, **38**: 663–665.

Singh, A.K., Choudhary, R.K. and Sharma, R.P.R. 1993b. Effect of inoculation and fertilizer levels on yield, nutrient uptake and economics of summer pulses. *Journal of potassium Research*, **9**: 176–178.

Singh, A.K. and Srivastava, S.K. 1985. Nodular physiology of urd bean as affected by urd bean mosaic virus–II: Effect on nitrogen content. *Indian Journal of Virology*, **1**: 325–329.

Singh, A.K. Srivastava, S.K. 1988. Interaction of *Rhizobium* and urd bean mosaic virus infection on the nitrogen metabolism of urd bean. *Phyton-Horn*, **28**: 161–169.

Singh, A.K. Srivastava, S.K. and Singh, S.V. 1985. Nodular physiology of urd bean affected by bean mosaic virus-VI. Leghaemoglobin content. *Acta Botanica Indica*, **13**: 131–133.

Singh, A.K. Srivastava, S.K., Singh, A. and Singh, M. 1986. Effect of urd bean mosaic virus and *Rhizobium* interaction on growth pattern of urd bean. *Indian Journal of Plant Pathology*, **4**: 116–120.

Singh, B.G. and Sudhakar, P. 1991. Effect of inoculation and phosphorus fertilization on dry matter production, N and P composition and yield of black gram. *Indian Journal of Plant Physiology*, **34**: 297–301.

Singh, M., Singh, J. and Singh, K. 1983. Effect of phosphorus and biofertilizers on chlorophyll content of leaves and leghaemoglobin content in fresh nodules in kharif grain legumes. *Indian Journal of Agronomy*, **28**: 229–234.

Singh, N., Yadava, H.S. and Dhakarey, R.P.S. 1993. Genetic variability for physiological parameters in urd bean. *Indian Journal of Pulses Research*, **6**: 96–98.

Singh, N., Yadava, H.S. and Dhakarey, R.P.S. 1994. Stability analysis for physiological parameters and seed yield in black gram. *Indian Journal of Plant Physiology*, **37**: 209–213.

Singh, N.P., Tagat, V. and Singh, S. 1987. Effect of excess water on growth and yield of black gram, *Vigna mungo* (L) Hepper. *Annals of Agricultural Research*, **8**: 74–80.

Singh, R. and Mall, T.P. 1974. Studies on the nodulation and nitrogen fixation by infected leguminous plants. I. Effect of arhar mosaic virus infection on nitrogen value, nodulation and nitrogen fixation by some pulse crops. *Plant and Soil*, **41**: 279–286.

Singh, R. and Mall, T.P. 1978. Response of nitrogen application on the growth, nodulation and nitrogen value of virus infected urd bean. *Science and Culture*, **44**: 550–552.

Singh, R.K. and Raghuvanshi, S.S. 1985. Pentaphyllous mutant in *Vigna mungo* (Urd). *Current Science*, **54**: 1286–1287.

Singh, R.K. and Raghuvanshi, S.S. 1987. A high yielding mutant of black gram with improved nodule formation. *Mutation Breeding News Letter*, **30**: 16–17.

Singh, S.B. and Verma, B.S. 1985. Effect of kharif grain legumes on nitrogen economy in succeeding crop of wheat. *Indian Journal of Agronomy*, **30**: 397–400.

Singh, T.B. 1989. Root nodule formation and leghaemoglobin concentration in some important legumes of North India. *Advances in Plant Sciences*, **2**: 155–158.

Singh, V. K. and Mathur, S.N. 1989. Studies with phorate, an organophosphate insecticide, on growth, nodulation and on some enzymes of nitrogen metabolism in black gram. *Indian Journal of Plant Physiology*, **32**: 314–319.

Singh, V.P., Dubey, A. and Srivastava, A. 1993. Radioprotective effect of gibberellic acid in black gram [*Vigna mungo* (L) Hepper]. *Journal of Indian Botanical Society*, **72**: 285–287.

Singh, V.P.N. and Uttam, S.K. 1994. Effect of nitrogen levels on yield of wheat grown after kharif maize and with and without bacterial nodulation. *Bharatiya Krishi Anusandhana Patrika*, **9**: 102–108.

Siripin, S., Murakami, T., Srinives, P., Boonkerd, N., Wadisirisak, P. and Thavarasook, C. 1991. Screening of black gram cultivars to enhance N_2 fixation. Proceedings of the mung bean 90, Mai, Thailand, February 23–24, 1990, pp 117–123.

Soam, S.K. and Agrawal, P.K. 1989. Effect of dimethoate on nodulation and nitrogenase activity of nodules in pea (*Pisum sativum*) and black gram (*Phaseolus mungo*). *Indian Journal of Agricultural Sciences*, **59**: 824–825.

Srivastava, G.P. and Srivastava, V.C. 1994. Effect of fertilizers (NPK) on yield and monetary return of rainfed pigeon pea/urd bean inter-cropping. *Journal of Research, Birsa Agricultural University*, **6**: 135–136.

Srivastava, R.C. and Mathur, S.N. 1980. Nodulation and nitrate reductase activity in nodules and in leaves of black gram (*Vigna mungo*) as affected by varying day lengths. *Indian Journal of Experimental Biology*, **18**: 300–302.

Srivastava, R.C. and Mathur, S.N. 1981. Effect of gibberellic acid on nitrate reductase activity in root nodules of *Phaseolus mungo* L. *Annals of Botany*, **47**: 147–149.

Subba Rao, N.S. and Tilak, K.V.B.R. 1977. Souvenir Bulletin, Directorate of Pulses Development, Govt. of India.

Subba Rao, N.S., Tilak, K.V.B.R. and Singh, C.S. 1980. Root nodulation studies in *Aeschynomene aspera*. *Plant and Soil*, 56: 491–494.

Sud, R.K. and Gupta, K.G. 1972. The growth promoting effect of TCA and Dalapon on *Rhizobium* strains. *Indian Journal of Microbiology*, **12**: 263.

Sudarsanam, S. and Prasad, N.N. 1972/1973) Studies on the performance of x-ray mutants of rhizobial strains isolated from black gram, green gram and horse gram. *Annamalai Univ., Agric. Ann.*, **4/5**: 69–74.

Sudhakar, P., Singh, B.G. and Rao, L.M. 1989. Effect of *Rhizobium* and phosphorus on growth parameters and yield of black gram (*Phaseolus mungo*). *Indian Journal of Agricultural Sciences*, **59**: 402–404.

Swaminathan, R. and Prasad, N.N. 1976. Effect of quinolphos and carbofuran on the nodulation of black gram. *Madras Agricultural Journal*, **63**: 641–643.

Tangcham, B., Wadisiriusk, P. Aroonsri, C. and Kato, K. 1996. Collection and use of tropical rhizobia for increasing legume yield in Thailand. The second Ministry of Agriculture, Forestry and Fisheries, Japan, International Work Shop on Genetic Resources: Soil Microorganisms, pp. 71–78.

Thakur, N.S. 1999. Effect of phosphorus and sulphur levels with and without *Rhizobium* inoculation on growth and yield of black gram (*Vigna mungo* L.). *Advances in Agricultural Research in India*, **12**: 67–70.

Thakur, R.C. and Negi, S.C. 1985. Effect of fertilizers and *Rhizobium* inoculation in black gram. *Indian Journal of Agronomy*, **30**: 501–504.

Tilak, K.V.B.R. 1991. Bacterial Biofertilizers. Published by I.C.A.R., New Delhi, pp. 66.

Tiwari, S.P. 1998. Interaction of *Heterodera cajani* and *Rhizoctonia bataticola* with *Vigna mungo*. *Annals of Plant Protection Sciences*, **6**: 33–36.

Tiwari, V.N., Lehri, L.K. and Pathak, A. N. 1989. *Rhizobium* inoculation of legumes as influenced by phosphorus and molybdenum fertilization. *Journal of the Indian Society of Soil Science*, **37**: 712–716.

Tripathi, M.L., Namdeo, K.N., Tiwari, K.P. and Kurmvanshi, S.M. 1994. Relative efficiency of nitrogen and *Rhizobium* inoculation on growth and yield of kharif pulses and oilseeds. *Crop Research*, **7**: 328–333.

Tuladhar, J.K. 1977. Survey of nodulation pattern of leguminous crops. *Biology and Fertility of Soils*, **20**: 275–283.

Udaiyan, K., Manian, S., Muthukumar, T. and Greep, S. 1995. Biostatic effect of fumigation and pesticide drenches on an endomycorhizhal — *Rhizobium* — legume tripartite association under field conditions. *Biology and Fertility of Soils*, **20**: 275–283.

Upadhyay, R.M., Singh, B. and Katiyar, S.K. 1991. Effect of nitrogen, phosphate and sulphur application to black gram on yield rate and fate of P in an inceptisol. *Journal of the Indian Society of Soil Science*, **39**: 298–301.

Vaishya, V.K. and Gajendragadkar, G.R. 1982. Effect of rhizobial inoculation on nodulation and yield of different genotypes of Urid [*Vigna mungo* (L) Hepper]. *Indian Journal of Microbiology*, **32**: 132–134.

Venkateswarlu, V., Rao, V.V., Rao, A.S. and Rao, D.S.K. 1988. Effect of herbicides and phosphorus levels on nodulation, phosphorus uptake and yield of urd bean. *Indian Journal of Pulses Research*, **1**: 70–72.

Verma, H.N., Rastogi, P., Prasad, V. and Srivastava, A. 1985. Possible control of natural virus infection of *Vigna radiata* and *Vigna mungo* by plant extracts. *Indian Journal of Plant Pathology*, **3**: 21–24.

Yadav, D.S. 1990. Studies on nitrogen economy through mung bean and urd bean legumes in legume — cereal wheat cropping system. *Legume Research*, **13**: 65–68.

Yadav, N.K. and Vyas, S.R. 1971. Response of root nodule rhizobia to saline, alkaline and acid conditions. *Indian Journal of Agricultural Sciences*, **41**: 875–881.

Yadav, R.L. 1981. Nodulation and N-fixation by legumes grown in intercrops of sugarcane and their effect on sugarcane production. *Madras Agricultural Journal*, **63**: 29–35.

Young, J.P.W. and Haukka, K.E. 1996. Diversity and phylogeny of rhizobia. *New Phytologist*, **133**: 87–94.

Zaman, A. and Choudhuri, S.K. 1999. Performance of winter crops as influenced by different preceding crops under rainfed upland condition of West Bengal. *Journal of Interacademicia*, **3**: 150–155.

CHAPTER 5

DIVERSITY IN SEXUAL AND POLLINATION SYSTEMS AND POLLINATORS OF ANGIOSPERMIC PLANTS OF INDIA

J.B. Atluri, S.P.Venkata Ramana and *C. Subba Reddi**

Introduction

Sexual reproduction in flowering plants is of selective advantage over other forms of reproduction, in that it facilitates gene recombination leading to the formation of new genotypes, thus paving the way to the evolution of genetic diversity and speciation. To use the expression of Stebbins (1950), "Sex exists .... for any (no) reason that its function in securing a great variety of genetic recombinations, by which the evolutionary line may adapt itself to new and varied environments". This benefit can be obtained by the plants provided they live in a spatially and temporally varying environment (Solbrig, 1976). Spatio-temporal variations in environmental features over a local scale are much more pronounced in the tropics than in the temperate zones, and such variations may be due to climatic, edaphic and biotic factors (Ashton, 1969). It is understandable then that the variety of plants in the tropical environments might utilize to the extent possible the evolutionary potentials open to them from sexual reproduction.

The sexual process involves two important events: Pollination and fertilization, which occur in sequence and lead to the formation of fruit

* Department of Botany, Environmental Sciences, Andhra University, Visakhapatnam-530003

and seed, i.e., the next generation, thus ensuring the continuity of species. However, parental contribution to genetic makeup of the next generation of species is dependent on its sex expression and the vectors of pollen or pollinators. A variety of sex expressions are reported to occur in flowering plants, and the term sexual (breeding) system deals with all aspects of sex expression in plants that affect the relative genetic contributions to the next generation of individuals within a species (Wyatt, 1983). Pollination or the transfer of pollen from the anthers of a stigma may take place within the same flower (autogamy), or between flowers within the same plant (geitonogamy), or between flowers from different conspecific plants (xenogamy). The term self-pollination refers to autogamy and cross-pollination to geitonogamy and xenogamy. Most plants rely on vectors or external agents for their pollination and it is generally agreed that angiospermic flowering plants have evolved and diversified, in large part, because of their reliance on insects as agents of pollination. Therefore, an understanding of various sexual systems and pollination strategies including pollen vectors may provide insights into the understanding of genetic diversity and speciation in angiosperms.

Sexual Systems

The available literature on the sexual systems in flowering plants is mind-boggling. The diversified sexual systems in higher plants were considered and discussed at length by Fryxell (1957), and cogent and lucid treatments of the same are rendered by Frankel and Galun (1977), Bawa and Beach (1981), and Richards (1986). Based on floral sexuality, whether bisexual or unisexual, the diversified plant species are classified into distinct classes of sexual systems such as hermaphrodite (flowers having both stamens and pistil), monoecious (staminate and pistillate flowers in one individual plant) and dioecious (staminate and pistillate flowers on separate individual plants). Combinations or permutations of the above have yielded other sexual mechanisms such as andromonoecy (plant with both hermaphrodite and staminate flowers), gynomonoecy (plant with both hermaphrodite and pistillate flowers), androdioecy (staminate and hermaphrodite plants in a population) and gynodioecy (pistillate and hermaphrodite plants in a population). These situations of sex expression aided by such outcrossing mechanisms of self-incompatibility (Richards, 1986), dichogamy (protandry and protogyny) (Bertin and Newmann, 1993), heterostyly (Bahadur, 1968; Ganders, 1979) and herkogamy regulate gene flow to varying levels from one generation to the other. Some authors analysed angiosperm floras of different geographical regions of the world for the relative frequency of different sexual systems (see Table 1). In any flora, the dominant component comprises of hermaphrodites, the lowest proportion of 63% in Mexico and the highest of 92% in British flora. Among the different life forms, the hermaphroditic sexual system occurs in a greater proportion in herbs (Ramirez and Brito, 1990). Roy's (1974) analysis of Indian flora based on Hooker's flora of British India gives the proportion of hermaphrodites as 79.42%. In the tropical communities, genetic self-incompatibility occurs in a high degree - 76% (Bullock, 1985), 79% (Bawa, 1974), 67% (Ruiz and Arroyo, 1978), and 86% (Bawa, *et al.*, 1985). In the following, the morphological as well as functional sexual systems observed in some Indian plants are presented (Table 2). The various plants are arranged on the sexual systems continuum with hermaphroditism at one extreme with possibilities for obligate selfing, and with dioecism at the other extreme as an obligate outcrossing mechanism. The various sexual systems observed are discussed in the light of related literary data from other countries. Only those categories of sexual systems in which the plant species investigated in India could be accommodated only considered.

Table 1: Distribution of Different Sexual Systems Among Species of Different Geographical Regions of the World, as Percentage of Total Angiosperms.

Sexual system	*India*	*World*	*Puerto Rico*	*Mexico*	*S.Africa*	*Britain*	*Venezuela*	*Costa Rica*
Hermaphrodite	79.42	72	78.9	63	77.7	92	77.34	65.5
Monoecious	13.86	5	10.5	9	2.6	5	19.53	11.4
Dioecious	6.72	4	6.1	27	6.6	2	3.13	23.1
Andromono-ecious		2	2.4		4			
Gynomon-oecious		3	2		7.1			
Heterostyly					1.9			

Source:
India — Roy (1974)
World — Yampolsky and Yampolsky (1922)
Puerto Rico — Flores and Schemske (1984)
Mexico — Ibarra-Manriquez and Oyama (1992)
S.Africa — Steiner (1988)
Britain — Lewis (1942)
Venezuela — Ramirez and Brito (1990)
Costa Rica — Bawa *et al.*, (1985)

Table 2: Sexual Systems in Some Indian Angiosperms

Plant species	*Morphological*	*Functional*	*Compatibility*		
			Auto	*Geitono*	*Xeno*
DICOTS					
Allophylus serratus	Andromonoecy	Monoecy		+	+
Bauhinia purpurea	Bisexual		+	+	+
B. racemosa	Dioecy				+
Cardiospermum halicacabum	Andromonoecy	Monoecy		+	+
Cassia alata	Bisexual			+	+
Chrozophora rottleri	Monoecy, PG			+	+
Cicca acida	Monoecy, PG			+	+
Clerodendrum infortunatum	Bisexual, PA			+	+
Cochlospermum religiosum	Bisexual				+
Coriandrum sativum	Andromonoecy, PA			+	+
Croton bonplandianum	Monoecy			+	+
Daucus carota	Andromonoecy, PA			+	+
Duranta repens	Bisexual			+	+
Emblica officinalis	Monoecy, PG			+	+
Euphorbia antiquorum	Monoecy, PG			+	+
E. dracunculoides	Monoecy, PG			+	+
E. geniculata	Monoecy, PG			+	+
E. tortilis	Monoecy, PG			+	+
Gloriosa superba	Bisexual, HK		+	+	+
Gmelina arborea	Bisexual				+
G. asiatica	Bisexual				+
Hedyotis nigricans	Distyly				+
Helicteres isora	Bisexual		+	+	+
Hugonia mystax	Distyly				+
Jatropha gossypiifolia	Monoecy, PG			+	+
Madhuca indica	Bisexual				+
Manilkara zapota	Bisexual		+	+	+
Martynia annua	Bisexual		+	+	+
Moringa oleifera	Bisexual			+	+
Oldenlandia scopulorum	Distyly				+
O. umbellata	Distyly				+
Phyllanthus pinnatus	Dioecy			+	+
Premna latifolia	Bisexual			+	+
Pterocarpus santalinus	Bisexual		+	+	+
Santalum album	Bisexual				+
Sapindus emarginatus	Andromonoecy	Temporal dioecy		+	+
Spathodea companulata	Bisexual				+
Sterculia foetida	Andromonoecy			+	+
Tamarindus indica	Bisexual		+	+	+
Tectona grandis	Bisexual		+	+	+
Terminalia catappa	Andromonoecy		+	+	+
T. tomentosa	Bisexual				+
Torilis arvensis	Andromonoecy, PA			+	+
T. leptophylla	Andromonoecy, PA			+	+

Contd...

Table 2 contd...

Plant species	*Morphological*	*Functional*	*Compatibility*		
			Auto	*Geitono*	*Xeno*
Vitex negundo	Bisexual		+		
Xeromphis spinosa	Dioecy				+
Zizyphus mauritiana	Bisexual	Temporal dioecy		+	+
Z. oenoplia	Bisexual	Temporal dioecy		+	+
GRASSES					
Brachiaria distachya	Bisexual, PA		+	+	+
B. mutica	Bisexual, PA				+
B. ramosa	Bisexual, PA		+	+	+
B. reptans	Bisexual, PA		+	+	
B. subquadripara	Bisexual, PA				+
Chionachne koenigii	Monoecy, PG				+
Chloris barbata	Bisexual, PA		+		
Chrysopogon aciculatus	Andromonoecy, PG				+
Cynodon dactylon	Bisexual, PA		+	+	+
Dactyloctenium aegyptium	Bisexual, PA		+	+	+
D. sindicum	Bisexual, PA		+	+	+
Dichanthium annulatum	Bisexual, PA		+	+	+
Digitaria ciliaris	Bisexual, PA		+	+	+
D. longiflora	Bisexual, PA		+	+	+
Echinochloa colonum	Bisexual, PA		+	+	+
E. crusgalli	Bisexual, PA		+	+	+
E. stagnina	Bisexual, PA				+
Eleusine indica	Bisexual, PA		+	+	+
Eragrostis riparia	Bisexual, PA		+	+	+
E. tenella	Bisexual, PA		+	+	+
Eriochloa procera	Bisexual, PA		+	+	+
Imperata cylindrica	Bisexual, PG				+
Iseilema laxum	Monoecy, PG				+
Leersia hexandra	Bisexual, PA		+	+	+
Leptochloa chinensis	Bisexual		+		
L. cynosuroides	Bisexual		+		
Oplismenus burmannii	Bisexual, PA				+
Panicum antidotale	Bisexual, PA		+	+	+
P. repens	Bisexual, PA		+	+	+
Paspalidium flavidum	Bisexual, PA		+	+	+
Paspalum orbiculare	Bisexual, PA		+		
Pennisetum pedicellatum	Andromonoecy, PG				+
Pseudobrachiaria deflexa	Bisexual, PA				+
Rhynchelytrum repens	Bisexual, PA		+	+	+
Saccharum spontaneum	Bisexual, PG				+
Setaria glauca	Bisexual, PA		+	+	+
Sorghum halepense	Bisexual, PA				+
Sporobolus tremulus	Bisexual, PG		+	+	+
Vetiveria zizanioides	Bisexual, PA				+

PG=Protogyny; PA=Protandry; HK=Herkogamy

Hermaphroditism

The flowers in the basal angiosperms are typically hermaphrodite (Thien *et al.*, 2000), suggesting that the first angiosperm flowers were probably hermaphrodite. It is thought that combination of pollen-producing and pollen-receiving organs in the same flower allows for efficient, simultaneous deposition and removal of pollen, and represent the optimal use of energetic resources available for reproduction. Hermaphrodite flowers are generally large and showy, and are present in a diverse group of plants. Hermaphrodite species may face some genetic problems since such sexual system permits self-pollination and fertilization to effectuate. But such possibilities may be circumvented by mechanisms of herkogamy, dichogamy, and self-incompatibility. However, these mechanisms are not simply interchangeable alternative 'solution' for avoiding self-pollination and fertilization. Some species may be adapting a dual mechanism of self-incompatibility and dichogamy (Lloyd and Yates, 1982). In such cases, dichogamy serves to avoid 'stigma clogging' with self-pollen.

Most plant species produce more flowers than they mature to fruit. The surplus flowers are thought to function as males and thus to enhance male fitness. A situation of this kind is reported in the Indian ber *Zizyphus mauritiana* and *Z. oenoplia* (Rama Devi *et al.*, 1989; Atluri *et al.*, 1995). In these species, the flowers are strongly protandrous, and flower life is sharply delineated into male and female stages. Those flowers that are produced later in the inflorescence and those inflorescences produced similarly in the plant usually stop with the male stage. Further, at the plant level two categories, one with early anthesis and the other with late anthesis are identified. The early and late phenotype segregation associated with strong protandry facilitates to a greater extent xenogamous fruit production. It is thought that Rhamnaceae are adapted to this strategy to probably maximise xenogamy.

The Verbenaceae species possess diversified functional sexual systems. *Tectona grandis* and *Vitex negundo* are compatible to auto, geitono and xeno pollen, but the former is predominantly xenogamous while the latter is superior in geitonogamy (Byragi Reddi and Subba Reddi, 1994). *Premna latifolia, Clerodendrum infortunatum* and *Duranta repens* are geitono and xeno compatible. *Gmelina arborea* and *G. asiatica* are also compatible to geitono and xenogamy, but fruit of xenogamy only reaches maturity, the other is abscised, thus making the species an obligately outcrosser (Subba Reddi *et al.*, 1996). A similar mechanism operates in *Cochlospermum religiosum* of Bixaceae (Bhaskara Rao and Subba Reddi, 1994). In the family Ceasalpiniaceae, *Bauhinia purpurea* and *Tamarindus indica* are compatible to auto, geitono and xeno pollen but with dominance of xenogamy. However, autogamy is totally precluded due to strong protandry; stigma attains receptivity a day after in *B. purpurea*. Erantiostyly has the same effect in *T. indica*. *Casssia alata* is compatible to geitono and xeno pollen, the latter being superior (Bhaskara Rao *et al.*, 1998). The Fabaceae member *Pterocarpus santalinus*, a species endemic to Andhra Pradesh and identified as endangered, is compatible to auto and geitono and xeno pollen, the last one being superior. This species is a mass bloomer with characteristic episodic flowering and opening its flowers at around midnight. Anthers dehisce successively over a two and half hour period, thus spreading pollen over a longer period and creating an opportunity for the visitors (*Apis dorsata*) to pick-up pollen on more occasions, and thus mostly promoting outcrossing (see Lloyd and Yates, 1982; Ramirez *et al.*, 1984). Such pollen presentation mechanism also reduces geitonogamous pollen discounting (Harder and Thomson, 1989; Harder and Barett, 1995). Large scale fruit abortion occurs. Probably the fruit from selfing is abscised (Rao *et al.*, 2001). In Sapotaceae, while *Madhuca indica* is obligately outcrossing (Kuruvilla, 1989), *Manilkara zapota* accepts auto, geitono and xeno pollen, with a preference for geitono pollen (Reddi, 1989). *Helicteres isora* of Sterculiaceae is compatible to auto, geitono and xeno pollen (Atluri *et al.*, 2000a). The Terminalias (*T. tomentosa, T. arjuna, T. chebula, T. bellarica, T. paniculata*) of Combretaceae (Srivastava, 1993), and the sandal wood tree *Santalum album* (Jyothi *et al.*, 1991) are all obligate outcrossers enforced by self-incompatibility. *T. tomentosa* is a mass bloomer. Strong protandry and staggered anther dehiscence avoid stigma clogging. Nectar resorption at around 1100 h drives the floral visitors to move enmasse to other conspecifics, thus achieving an effective delivery of cross pollen (Atluri *et al.*, 2000b). *Spathodea companulata* is also self-incompatibility and outcrossing. *Ceiba pentandra* is compatible to auto, geitono and xeno pollen, the latter being superior (Subba Reddi *et al.*, 1989). *Moringa oleifera* is compatible to geitono and xeno pollen, the latter performing well (Jyothi *et al.*, 1990). *Martynia annua* gives 100% fruit through autogamy, and fruiting through other modes is less than 50%. Autogamy with a provision for xenogamy is advantageous to this weedy species, enabling it to colonise many different niches in disturbed habitats (see Allard, 1965; Mulligan, 1972). The endangered species *Gloriosa superba* of Liliaceae displays herkogamy but this is mostly wind-pollinated (Subba Reddi *et al.*, 1998). *Hugonia mystax* is distylous, with 'pin' and 'thrum' morphs occurring in equal frequency (Raju *et al.*, 1997). Plants display episodic and steady state flowering, which are known to promote outcrossing.

Analysis of a functional sexual system in 35 morphologically hermaphrodite grass species revealed that nine of these species are predominantly outcrossing through the operation of self-incompatibility, another five are predominantly self-compatible and the remaining 21 species are equally compatible to self and cross pollinations (Table 2). The last group of 21 species with a balanced sexual system may also be considered to go for outcrossing in preference to self provided the stigmas received self- and cross-pollen simultaneously. In such situations of competition between cross- and self-pollen, it is assumed that cross-pollen is always competitively superior to self-pollen, and therefore cross-pollen is always favoured by the maternal tissue in the pistil. Competition results in the better growth of cross-pollen, which are allowed to enter the ovary. This kind of self-incompatibility triggered by the cross- and self-pollen competition is termed 'cryptic' self-incompatibility by Bateman (1956). Subsequently this system has been reported in *Ansinckia glandiflora* (Weller and Ornduff, 1977) and *Clarkia ungulata* (Bowmann, 1987). The concept of cryptic self-incompatibility is adopted by Hessing (1988) to explain the dilemma of geitonogamous pollinations in *Geranium caespitosum*, where manually geitonogamously pollinated flowers initiated as many fruits as manually outcrossed flowers, but showed a subsequently greater number of embryo abortions. Bowmann (1987) suggests that 'cryptic' self-incompatibility can enforce outcrossing without eliminating the possibility of selfing in the event of cross-pollen not reaching the stigma. Weller and Ornduff (1977) predicts that this system may be more widespread in angiosperms. In the tropical tree species *Shorea robusta*, strong protogyny expressed well before the unfolding of petals protects the stigma from auto-pollen deposition and permits deposition of geitono and/or xeno pollen which may result in outcrossing through 'cryptic' self-incompatibility (Atluri *et al.*, 2003). Such interpretation agrees with the argument that long-lived species like the sal tree should strongly favour outcrossing (Ledig, 1986).

Andromonoecism

This sexual system is rather common among angiosperms. It is considered to have evolved from hermaphroditism by the loss of female reproductive structures from a proportion of plant's flowers (Lewis, 1942), and generally assumed to be a mechanism for outcrossing. But other mechanisms are often cited promoting outcrossing in andromonoecious species. Such mechanisms are self-incompatibility, dichogamy and sequential maturation of different flower types on the same plant. The increase in number of polleniferous flowers in andromonoecious species has other advantages. It results in conspicuous floral display, thus serving to attract pollinating visitors (Primack and Lloyd, 1980). It is more a mechanism for adjusting reproductive effort to male or female functions (May and Spears, 1988), and increases male fitness since the resources saved by producing fewer pistils could be directed to male function (Emms, 1993). Also packing the pollen into many small units rather than a few large ones enhances pollination success by facilitating pollen presentation over an extended period (Harder and Thomson, 1989; Emms, 1993).

In the andromonoecious species, possibilities are open up for auto, geitono and xenogamy. These possibilities are examined in some andromonoecious species. Sequential opening of florets in the grass species *Chrysopogon aciculatus* indicated that bisexual spikelets are protogynous, those opened up on the first day of the flowering life of the first whorl in a spike are exposed to xenogamous pollen on the day of their opening only. But on the second day, its own pollen is available creating a possibility for autogamy. The pollen of the bisexual spikelet in the first whorl is also available to the females of the second and third whorls in which geitonogamy is likely to take place. If geitonogamy fails, their own pollen is available the following day for autogamy to occur (Table 3). In spite of the possibilities for auto and geitono selfing, the species has a low level of self-compatibility and a high level of xenogamy. Obviously reproduction is dependent on outcrossing enforced by self-incompatibility. Another andromonoecious grass species *Pennisetum pedicillatum* is strongly protogynous with two days of gap between maturation of essential organs in bisexual floret; the male florets ripen after a gap of five days in one spikelet Stigma is receptive for two days and no auto-pollen is available during that receptive period. Possibilities are open up only for geitono and xenogamy. However, the quality of seed obtained through geitonogamy is very low (25% germinability) compared to xenogamy (92% germinability), thus revealing a greater compatibility for xenogamous pollen.

Sapindus emarginatus, Allophylus serratus, Cardiospermum halicacabum and *Sterculia foetida* are all andromonoecious. In these species anthers in the bisexual flower are non functional, making the plants functionally monoecious. *S. emarginatus* and *A. serratus* display characteristic flowering phenology with alternation of staminate and pistillate functions within the individual plants. Majority of the individual plants have first the staminate phase followed by pistillate, and again the staminate phase. The gap between different phases is 1–3 days. Such a phenotypic pattern with alternation of sexual

Table 3: Sequential Maturation of Bisexual and Male Spikelets Showing the Stigma Receptivity Period and the Pollen-Flow Directionality in *Chrysopogon aciculatus*.

Day	First whorl			Second and third whorls			All other whorls		
	Bisexual spikelet		Male spikelet	Bisexual spikelet		Male Spikelet	Bisexual spikelet		Male spikelet
	Female	Male		Female	Male		Female	Male	
1	FP ® 6	0		0	0		0	0	
2		6		18	0		0	0	
	↓	←	——→						
3				0	18		0	0	
				↓	←				
4							FP ® 24	0	
5							0	24	
							↓	←	
6									
7			12						
8						36			
9									
10									48

FP - Foreign pollen
Arrows indicate pollen-flow directionality

functions is described as duodichogamy. This pattern is interpreted to have evolved in response to selection pressure for outcrossing. Cruden and Hermann-Parker (1977) coined the term 'temporal dioecism' to describe this pattern of sexual phase alternation within the individual. A synonym for temporal dioecism is synchronous dichogamy (Lloyd and Webb, 1986). This phenotype undoubtedly enforces outcrossing provided the different phases do not overlap. However, some individuals have overlapping of the sexual phases in some branches or within individuals, thus opening up possibilities for geitonogamy. The plants are self-compatible through geitonogamy. Cruden and Hermann-Parker (1977), and Cruden (1988) proposed 'temporal dioecism' as a sexual system comparable to heterostyly and dioecy in facilitating outcrossing through the promotion of xenogamy. *C. helicacabum*, though taxonomically related to *S. emarginatus* and *A. serratus* — all belonging to the plant family Sapindaceae, does not display alternation of sexual phases. On any single occasion both staminate and pistillate flowers are available on the plant. This simultaneous occurrence of the two sexual phases allows self-pollination through geitonogamy, which gave 100% fruit in controlled pollinations. Selfing may be of a special adaptive value for this herbaceous weedy twiner species. However, this species is also compatible to outcross pollen, thus having the scope for maintaining genetic heterogeneity. In *Sterculia foetida*, the bisexual flowers become functionally pistillate due to male sterility. Thus the individual trees are functionally monoecious with staminate and pistillate flowers occurring simultaneously. Both flower morphs mature simultaneously on a day opening up a possibility for geitonogamy to occur. But the pistils are incompatible to selfing through geitonogamy. Thus the system enforces total outcrossing by mitigating the staminate function of hermaphrodite flowers. *Terminalia catappa* is andromonoecious with hermaphroditic flowers at the base of the inflorescence and the staminate flowers above, contra other *Terminalia* species described under hermaphrodites. Inflorescence length differs in different plants, and on the whole the plants could be categorized into three groups: one with long, another with medium and the other with short inflorescences. The anthers of both flower morphs of short inflorescences are abortive; the flowers produced normal fruit almost with 100% success through apomixis. Bisexual flowers of other categories of plants produced fruit through auto- and geitonogamy, but are totally incompatible to xenogamy. Thus it is evident that each of the andromonoecious species described above have their own peculiarities and adaptations to achieve outcrossing.

Monoecism

Monoecy is considered to be an advanced sexual system over andromonoecy and gynomonoecy (Bertin, 1989). It occurs in wind-pollinated species of the temperate region and in zoophilous species of the tropical forest. While the evolution of monoecy is generally attributed to selection for avoidance of self-fertilization, several workers have challenged this explanation, both by suggesting other possible advantages to this trait, and by pointing out the presence of self-incompatibility in monoecious species (see Bertin, 1989). Other possible advantages mostly attributed to monoecy are the avoidance of pollen-pistil interference and provision of greater flexibility in temporal and spatial patterning of resource allocation to male and female functions (Bawa and Beach, 1981;. Willson, 1983; Bertin, 1989). These two advantages select for the sterilization and abortion of stamens in hermaphroditic flowers of andromonoecy, thus giving way to the evolution of monoecy. But in some situations when stamens function in attracting the pollinators and/or providing pollen as the sole reward to the pollinators, the pollinators may select against the sterilisation or abortion of stamens in hermaphroditic flowers of andromonoecious species (see Bawa and Beach, 1981). But in cases different from the above, sterilisation of stamens of hermaphroditic flowers may occur making the species monoecious as is the case with *Sterculia foetida*. This species is morphological by andromonoecious, becoming monoecious through the nonfunctioning of stigmas. It proceeded further to become an obligate outcrosser due to incompatibility to geitonogamy. In the family Euphorbiaceae, *Emblica officinalis, Cicca acida, Euphorbia antiquorum, E. tortilis, E. geniculata, E. drancunculoides, Jatropha gossypiifolia, Croton bonplandianum* and *Chrozophora rottleri* are all monoecious. All are protogynous except *C. rottleri*. All are compatible to geitono and xeno pollen, and excepting the initial pistillate flowers that open, all others are likely to receive geitono and xeno pollen, and there is no device ensuring outcrossing advantage attributed to monoecy. But in *E. antiquorum* and *E. tortilis*, selfing through geitonogamy is avoided by temporal dioecism described earlier with reference to *Sapindus emarginatus* under andromonoecy. In these two species, the central cyathium in an apparently simple cyme matures first. This is functionally always staminate. A day or two following the cessation of staminate phase, the lateral cyathia mature; these are morphologically and functionally hermaphrodite with a strong protogyny. A day after the termination of pistillate phase, the second staminate phase begins. Duodichogamy is synchronous in an individual, and so at a point of time the plant

Table 4 : Maturation Sequence of Female and Male Spikelets in *Chionachne koenigii* with Monoecy and Protogyny.

Day	*Raceme (1 female, 8–14 male)*		*Node (8–20 racemes)*		*Culm (10–14 nodes)*	
	Female	*Male*	*Female*	*Male*	*Female*	*Male*
1	1	0	4	0	20	0
2	0	0	2	0	10	0
3	0	0	0	0	0	0
4	0	1	0	10	0	145
5	0	3	4	14	28	169
6	0	3	0	18	0	180
7	0	1	0	22	18	190
8	0	1	2	16	10	150
9			0	12	32	240
10			0	14	0	190
11			2	0	10	110
12			0	0	0	0
13			0	0	40	0
14			2	0	12	0
15			0	1	0	24
16			0	4	0	16
17			2	14	20	36
18			0	16	0	120
19			0	8	0	30
20			0	4	10	70
21			0	6	0	10
22					0	20
23					0	80

is either staminate or pistillate and thus there is no opportunity for geitonogamy to occur. In the case of *E. tortilis* some overlapping of sexual phases occur in some branches, with a possibility for geitonogamy. *Bridelia retusa* is also monoecious displaying temporal dioecy, but its operating sexual system is yet to be worked out (Borges *et al.*, 1997). The Umbellifer *Torilis leptophylla* is strongly protandrous at the flower level, but in the full plant pistillate and staminate phases overlap, opening up for a blend of geitono and xenogamy (Koul *et al.*, 1984).

Two grass species *Chionachne koenigii* and *Iseilema laxum* are monoecious with a strong protogyny. In a raceme of *C. koenigii,* the first spikelet that opens is the lone female, its receptivity lasting three days. From the fourth day onwards, the male spikelets begin to mature. The male spikelets varying between 8–14 mature in succession over a five day period. A time gap between female and male maturity occurs even at the node and culm level. Therefore, the first female spikelets that open up at the beginning of flowering of an individual receive

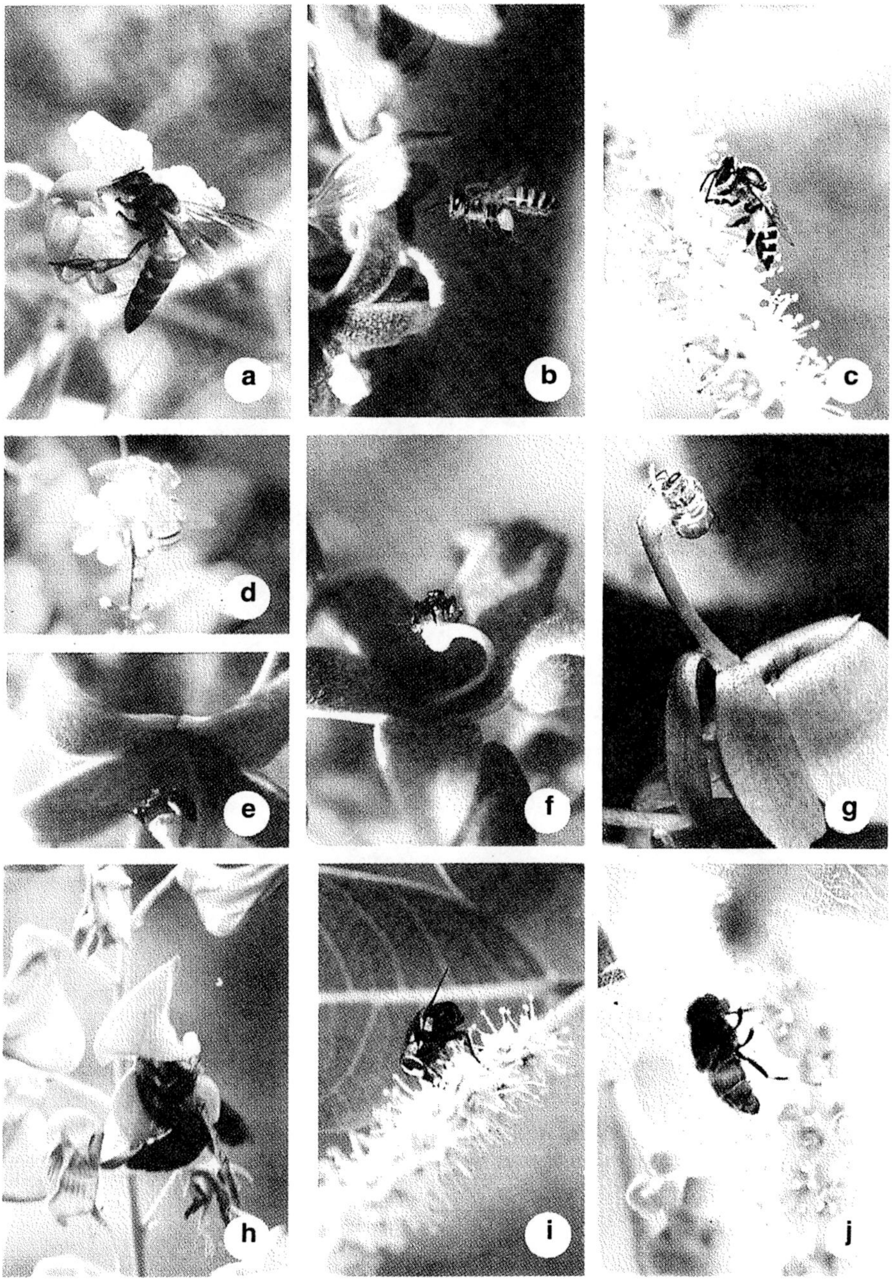

PLATE 1

Photographs of pollinators: a. *Apis dorsata*, b. *A. indica*, c. *A. florea*, d. *Amegilla* sp., e. *Cerantina* sp. f. *Trigona iridipennis*, g. *Pseudapis* sp., h. *Xylocopa latipes*, i. *Chrysomya megacephala*, j. *Eristalis arvorum*

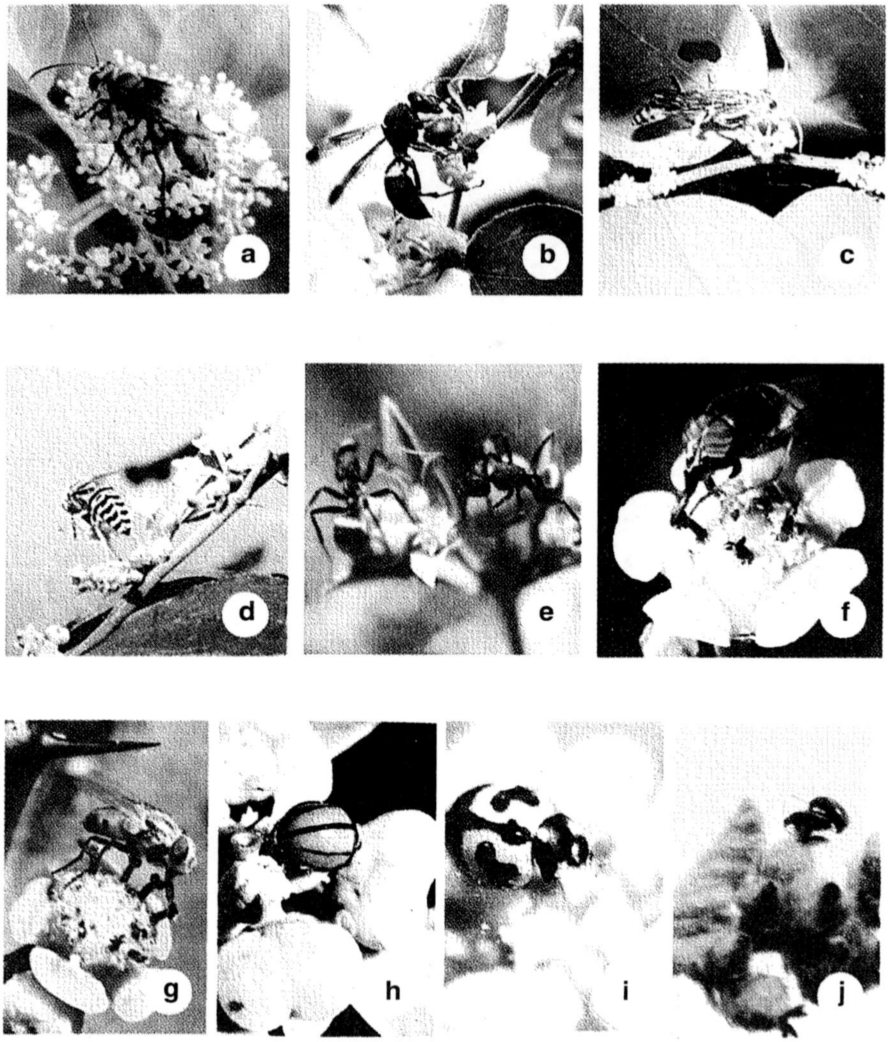

PLATE 2

Photographs of pollinators: a. *Sphex* sp., b. *Ropalidia spatulata*; c. *Chalybian* sp., d. *Bembix* sp., e. *Camponotus* sp., f. *Rhyncomya viridaurea*, g. Scarabaeidae beetle, h. *Varania discolor*, i. *Coccinella repanda*, j. Scarabaeidae bettles.

PLATE 3

Photographs of pollinators: a. *Sarcophaga* sp., b. *Musca domestica*, c. *Rhyncomya viridaurea*, d. Bombylid fly, e. Noctuid moth, f. Hawkmoth *Macroglossum gyrans*, g. *Turdoides affinis*, h. *Nectarinia zeylonica*, i. *Camponotus* sp.

PLATE 4

Photographs of pollinators: a. *Junonia almana*, b. *Hypolimnas bolina*, c. *Castalius rosimon*, d. *Everes parrhasius*, e. *Gangara thyrsis*, f. *Anaphaeis aurota*, g. *Catapsilla pomona*, h. C. *Pomona catilla* form

PLATE 5

Photographs of pollinators: a. *Junonia lemonias*, b. *J. Hierta*, c. *Phalanta phalantha*, d. *Hypolimnas missippus*, e. *Danaus chrysippus*, f. *Euploea core*, g. *Spindasis vulcanus*, h. *Papala jarbus*

PLATE 6

Photographs of pollinators: a. *Valeria valeria*, b. *Pachliopta hector*, c. *P. hector*, d. *Papilio polymnestor*, e. *Graphium agamemnon*

xeno-pollen. The other female spikelets that open up in succession are likely to receive both geitono and xeno-pollen (Table 4). In *I. laxum*, the two male spikelets in the raceme mature on the sixth day after female spikelet has matured. At the node level there are 3–8 racemes with a synchronous ripening of female and male spikelets. The female spikelets open up over a period of 1–4 days and the male ones over 4 days. At the culm level there are more number of female and male spikelets and in the first five days only females ripen. From sixth day onwards both mature females and males are available on the same culm. Stigma receptivity lasts for three days, hence the female spikelets that mature on the first three days on the culm are exposed for xeno-pollen and those opening subsequently receive both geitono-and xeno-pollen (Table 5). But the seed quality assessed in a germination study revealed the failure of geitonogamy producing viable seed. It is thus evident that these two monoecious grass species are obligately outcrossing enforced by incompatibility to geitonogamy.

Dioecism

Dioecy, separate genders, is widespread in angiosperms. It is more common in tropical floras than in temperate ones (Bawa, 1980; Ibarra-Manriquez and Oyama, 1992). This sexual system has evolved from hermaphroditic ancestors independently many times and possibly from different regions (Renner and Ricklefs, 1995). Two contrasting views are in sway with regard to the causal or selective factors behind the evolution of dioecy. The traditional view holds that selection for outcrossing or inbreeding avoidance is a sine qua non for the evolution of dioecy (see Thomson and Barrett, 1981), and the genetic models of Charlesworth and Charlesworth (1978) add strength to this explanation. Co-occurence of dioecy and self-incompatibility systems is cited to support this genetic model for the evolution of dioecy (Thomson and Barrett, 1981). Alternatively, sexual selection and selection for optimal resource allocation between the two genders are proposed independent of genetic considerations. Willson (1979), Bawa (1980) and Givinish (1980, 1982) provide the necessary evidence for the alternative ecological model showing differential male and female costs and success rates, and effective distribution of resources between male and female functions of dioecy. Certain correlates of dioecy like small, unspecialised flowers pollinated by small, unspecialised bees and flies, and fleshy fruits are also cited in support of the ecological model. Baker (1984) and Fox (1985) considered the two opposing models — genetic and ecological, and stated that these models are not mutually exclusive, thus both models may be effecting the evolution of

Table 5: Maturation Sequence of Female and Male Spikelets in *Iseilema laxum* with Monoecy and Protogyny.

Day	Involucral raceme (1 female, 2 male, 3 empty)		Node (3–8 involucral racemes)		Culm (4–8 nodes)	
	Female	Male	Female	Male	Female	Male
1	1	0	2	0	2	0
2	0	0	1	0	2	0
3	0	0	0	0	3	0
4	0	0	2	0	5	0
5	0	0	0	0	2	0
6	0	2	0	4	3	4
7			0	2	3	4
8			0	0	2	6
9			0	4	3	10
10					4	4
11					3	6
12					4	6
13					1	4
14					1	6
15					0	8
16					0	6
17					0	8
18					0	2
19					0	2
20					0	0

dioecy. An elaborative and logical discussion of the relative roles of sexual selection and inbreeding avoidance factors in the evolution of dioecy presents two contrasting pathways of the evolution of dioecy, (1) from hermaphroditism via monoecy and (2) from hermaphroditism via gynodioecy (Freeman *et al.*, 1997). Sexual specialization is attributed as the causal factor for the evolution of dioecy from close cosexual relatives possessing some means of promoting outcrossing, and this is applicable for the dioecious plants of monoecy pathway. Inbreeding avoidance is invoked as the selective factor behind the evolution of dioecy from hermaphrodites having no outcrossing mechanisms and these species followed the gynodioecy pathway. The first category of dioecious plants are normally wind-pollinated, can change sex in response to environmental stresses, the males predominating in xeric habitats and the females in mesic habitats

(Freeman *et al.*, 1981). In the second category of dioecious plants, the two sexes are genetically determined and the sex morphs are constant. Pollen dispersal may be either biotic or abiotic in both types of dioecy (see Freeman *et al.*, 1997).

Three of the plants in our study are dioecious; These are *Phyllanthus pinnatus*, *Bauhinia racemosa* and *Xeromphis spinosa*. Sexual difference in *X. spinosa* is revealed by stigma and anther characters. The pistillate flowers bear bi- or trilobed stigma, the lobes expanding at maturity, and the anthers are sterile. The staminate flowers contain unopened bilobed stigma and fertile anthers. The flowers are relatively large with nectar in quantity. Opened flowers are more numerous at dawn and dusk, and are foraged and pollinated solely by the diurnal hawkmoth *Macroglossom gyrans*. These floral traits are in deviation from the postulated traits of dioecious taxa of the tropics commonly known as Bawa hypothesis — small unspecialised flowers pollinated by unspecialized insects. This species occurs in xeric habitats (scrub jungles) but does not show any intersexual difference and the sex ratio is at unity, contra Freeman *et al.* (1997). *Phyllanthus pinnatus* has small and inconspicuous flowers, the anthers producing good amount of pollen which can be wind dispersed. The flowers also contain traces of nectar, which is being utilised by the small bees, flies and ants and possibly playing a role in pollination. Similarly the floral features of *Bauhinia racemosa* suggest wind and insect pollination. These two taxa also occur in xeric habitats and there is no unusual biasing of sex towards male.

Pollen Vectors/Pollinators

Pollination within a flower (autogamy) may be spontaneous (autopollination) or mediated by a vector. But geitonogamy and xenogamy occur only through the mediation of a vector, which includes water, wind and animals. The available information on pollinators at the global level was synthesized and packages of characters in plants pollinated by different vectors have been described in the form of pollination syndromes (Proctor and Yeo, 1972; Proctor *et al.*, 1996; Faegri and Pijl, 1979). Such grouping of characters is necessary for analyzing the chaotic diversity of flower types, but it over simplifies the complexity of functional pollination syndromes. Most of the plant species in the tropics are visited and pollinated by more than one biotic species often belonging to different orders or different classes, sometimes by insects and wind also, thus often leading to the existence of pollination ecotypes (Baker, 1985). Such floral adaptations are advantageous to the plant in heterogeneous environments as in the tropics where pollinator (biotic) availability is rather unpredictable. In such circumstances, it is important to quantify the relative contributions of different vectors as pollinators to a plant's sexual reproduction, and the principle of 'the most effective pollinator' has become central to discussions of plant-pollinator relationships (Stebbins, 1970). It is only by differential contribution to the plants' reproductive success, the pollen vectors influence the evolution of floral traits. Niklas (1985) analysed the theory and methodology of aerodynamics as they relate to the understanding of wind-pollination, and thus giving the background necessary for use in investigating the anemophilous plant species. Davis (1997) reviewed the efforts made by various researchers in evaluating animal visitors as pollinators and suggested that deposition of the amount of pollen withdrawn from a virgin flower during a single visit as well as the proportion of removed pollen that ultimately is deposited on conspecific stigma of recipient flower will provide a direct and effective measure of an animal's worth as a pollinator to the plant. In the following an attempt is made to ensemble the available data on pollen vectors of different plant species so far investigated among the naturally occurring angiospermic flora in India.

Autopollination

In some plant species, transfer of pollen within the same flower takes place with no external pollinating agent touching the flower. In some Asteraceae, if cross-pollination fails, the bifid stigmas bend and curve backwards so that their receptive surfaces come into contact with the pollen on the anthers to secure pollination. In some Malvaceae as observed in *Abutilon indicum* with flowers opening at around 1300 h, the petals close in the evening around 1800 h facilitating the pollen on the anthers and stigma coming together. In *Boerhavia diffusa* the flowers are rarely visited by insects. They achieve pollination by the coiling of style and stamen filaments. This is a regular feature taking place at around 0800 h in the morning (Chaturvedi, 1989). Autopollination may also occur due to the pollen falling down from the anthers on to the stigma, a mechanism described as gravity pollination. The *in situ* germination pollen of *Utricularia* species get liberated and deposited enmasse into the funnel shaped stigma by the inward bending of stamens and the dehiscence of anthers resulting in gravity pollination (Doreswamy and Mohan Ram 1969; Kausik and Raju, 1955). Spontaneous autogamy may be more frequent in the weedy plant species.

Water-Pollination

While reviewing the pollination mechanisms in Hydrocharitaceae, Cook (1982) described the floral mechanism in the species of *Elodea, Nechamandra, Vallisneria, Enhalus, Lagarosiphon* and *Appertiella* with adaptations for pollination at the water surface (Ephydrophily), and in the species of *Thalassia* and *Halophila* in the water (Hyphydrophily). Ghosh (1964) has described pollination in the monotypic genus *Nechamandra* which is native to Sri Lanka, South and Eastern India and Bangladesh. The male flowers from the spathes are released throughout the day but open in sunlight during 1100–1500 h. Each flower has 2 stamens which become horizontal and the pollen grains remain attached to the distal part of the filament. In the female plants, the flower bud breaks through the spathe and reaches the water surface. The sepals spread on the water surface, and the flattened stigmas arch towards the periphery of the flower. The sepals of both male and female flowers are water repellent and thus attract each other resulting in pollination. A similar mechanism operates in *Vallisneria,* except that the stamens remain upright and partially divergent. The male flowers are very small and are released on to the water surface. The perianth segments are unwettable. The calyx lobes are boat-shaped, and the stigmatic surfaces face upwards. The floating male flowers slide down on to the boat shaped calyx. There is a positive attraction between female and male flowers and the anthers are attached to the stigma.

Wind-Pollination

Pollination by wind is common in grasses and sedges, but rare in tropical trees. However, recent studies have proved and established the role of wind as a pollen vector of *Madhuca indica* (Subba Reddi, 1976; Kuruvilla 1989), *Mimusops elengi* (Subba Reddi and Janaki Bai, 1981), *Cicca acida* and *Emblica officinalis* (Subba Reddi and Reddi, 1984) and *Shorea robusta* (Atluri *et al.*, 2003). Although the floral morphologies of these taxa appear to be ambiguous with regard to their functional role in pollination and do not exhibit all the traditional traits of anemophily, they have certainly evolved towards utilizing the energy of wind for pollen transport. *Cicca acida* and *Emblica officinalis* deviate from the condition of shedding pollen under turbulent conditions for effective pollen transport. Pollen release occurs at 2000–0200 h in *C. acida* and at 1600–2100 h in *E. officinalis*. Turbulence by that time would be tending to decrease commensurate with decrease in temperature and increase in relative humidity. These two tree species do not occur in high densities; therefore, pollen shedding under such conditions not suitable for rapid pollen transport should be advantageous and help maintaining high levels of pollen concentration at the source for maximizing pollination. Similar deviations in the timing of pollen release also occur in grasses (Subba Reddi *et al.*, 1988). Majority of the grasses shed pollen during 0200–0800 h. During that time low turbulent conditions normally prevail permitting high pollen concentrations near the source for maximization of pollination. Since prevention of wastage of pollen is a legitimate consideration in the evolution of pollination, pollen shedding under conditions of low dispersive power of the atmosphere is in line with this principle of pollination economy, and appears to be an ideal strategy in view of the short longevity of grass pollen, 1–3 h in most cases. Yet another interesting mechanism occurs in *Shorea robusta*, a tropical tree species. The mature flower buds open up at any time of the day but with a higher frequency at 0800–1000 h and again at 1400–1600 h. Anther dehiscence occurs soon after the opening of flowers but no pollen release occurs. When shaken by the wind force there occurs explosive pollen release. This explosive release of pollen prevents wastage of pollen by simple gravitational fall from the pendulous flowers and ensures entrapment in moving air and further transport, thus satisfying an important condition for the efficiency of wind-pollination (Niklas, 1985). The process of wind-pollination is generally treated as a short distance phenomenon and the time gap between pollen release and pollen reception is anticipated to be short. However, source height influences transport distance, the distance increasing with the height of source. Pollen longevity of *S. robusta* extending over 50 h is in favour of this possibility. Explosive pollen release, prolonged pollen viability and tallness of trees may facilitate a relatively long distance pollen transport. Also the pollen grains conform to the typical aerodynamic particle size of 20–60 μm. Additional adaptations that contribute to the efficiency of wind-pollination in *S. robusta* include flowering before full leafing, gregarious occurrence permitting a high density of conspecifics, and high pollen production per flower (ca. 60,000 grains). These proven and confirmed cases of wind-pollination of tropical trees suggest that this mode of pollination in tropical forest trees is not as constrained as suggested in earlier writings of Whitehead (1969). The dioecious *Ailanthus excelsa* and *Borassus flabellifer* are also probably wind-pollinated.

Animal-Pollination

Different anthophilous animal groups recorded on some of the Indian plant species are set out in Table 6, and the diversified insect pollinator species are given in Table 7. The quantitative data were obtained through censuses in three phases of flowering: initial, peak and final, and the percentages given are based on the totals of the three phases. The relative importance of these anthophiles was determined on the basis of the criteria described and discussed in the preceding para. The photographs of some of the biotic pollinators are given in Plates 1–6. In the following, each of the different groups of anthophiles are considered, and their role as pollinators presented based on Indian studies. Wherever necessary relevant literature was borrowed from studies abroad, mostly from tropics.

Thrips

The association of thrips (Thysanoptera) with flowers is well established. They are known mostly as flower visitors and dwellers. There are only few reports disclosing their role as pollinators: European Ericaceae and Malaysian dipterocarps (see Kevan and Baker, 1999). In India, Mathur and Mohan Ram (1978) observed *Thrips hawaiiensis*, and *Haplothrips tenuipennis* inside the flowers of *Lantana camara* (Verbenaceae). The flower colour changes after pollination from yellow to shades of red, scarlet and mauve. The thrips utilised pollen grains and stigmatic exudates as food, and are associated only with the yellow phase of the flowers. The stigmas of yellow flowers only represent the receptive stage. These two species of thrips are also the pollinators of *Manilkara zapota* the Sapodilla fruit tree (Reddi, 1989). Thrips (*T. hawaiiensis* and *H. ceylonicus*) are implicated in the pollination of the avenue tree *Peltophorum inerme* (Mondal *et al.*, 1993).

While analysing in detail the adaptive significance of thrips with Asteraceae (Compositae) flowers, Ananthakrishnan (1982; 1993) mentioned about inter- and intra-specific competitions resulting in thrips migration, thus resulting in increased possibilities of cross-pollination. The report of these competitive interactions are particularly important and may weaken Mani's (1993) assumption that thrips as pollinators of Asteraceae are insignificant because they hardly ever fly out far from one inflorescence head to another. Many species of thrips are associated with Asteraceae and there is species-specific relation since none of the species are found with all the Asteraceae floral species reported (Ananthakrishnan, 1982). It is likely that in many plant species pollination by thrips may go undetected because simple exclusion experiments do not differentiate between autonomous pollination and thrips pollination.

Table 6 : Relative Abundance of Different Anthophilous Animal Groups as Percentage of Total Visits on Different Plant Species.

Flower Species	*Flower Visitor/Pollinator Groups*								
	Beetles	*Flies*	*Ants*	*Wasps*	*Bees*	*Moths*	*Butterflies*	*Birds*	*Bats*
Acanthus ilicifolius					75			25	
Allophylus serratus				34	64		**2**		
Anisomeles indica					67			33	
A. malabarica					69			31	
Bauhinia purpurea				**1**	97		**2**		
Brugeira sp.								100	
Cardiospermum helicacabum		**2**		**14**	83		**1**		
Cassia alata					100				
Ceiba pentandra		**1**		**1**	**15**				83
Chrozophora rottleri	95	**1**		**1**	**3**				
Clerodendrum infortunatum					**65**	**2**	33		
Cochlospermum religiosum					100				
Duranta repens				**1**	34	**14**	51		
Euphorbia antiquorum		95		**2**	**1**		**2**		
E. dracunculoides		**2**	94	**1**	**3**				
E. geniculata	7	3	56	27	7				
E. tortilis		25		**15**	60				
Helicteres isora				**6**	62		**9**	23	
Hugonia mystax					97		**3**		
Ipomoea kentrokaulos					100				
Jatropha gossypiifolia		8		8	76		**8**		
Leonotis nepetifolia					40			60	
Moringa oleifera					94	**1**	**5**		
Premna latifolia				75	**5**		20		
Pterocarpus santalinus					**100**				
Santalum album		7		35	56		**2**		
Sterculia foetida		48		**5**	**39**		**8**		
Tamarindus indica		**1**		**6**	92		**1**		
Tectona grandis		19		19	22	**2**	38		
Terminalia catappa		48	**0.5**		51		**0.5**		
T. tomentosa		18		12	58		**12**		
Tribulus terrestris	3	17			75		**5**		
Vitex negundo				43	40		**17**		
Xeromphis spinosa						100			
Zizyphus mauritiana		20	**2**	10	68				
Z. oenoplia		**13**	**30**	29	27		1		

Note: Numbers in bold indicate the effective pollinator group.

Table 7: Insect Pollinators of Indian Angiosperms.

Thysanoptera
(Thrips): Thripidae–*Frankliniella schultzei, Haplothrips goideyi, H. tardus, Megalurothrips distalis, Micothrips fasciatus, Microcephalothrips abdominalis, Thrips hawaiiensis.*

Coleoptera
(Beetles): Bruchidae–*Spermaphagus* sp.; **Chrysomelidae**–*Aulacophora lewisii;* **Coccinellidae**–*Coccinella repanda, Menochilus sexmaculata, Verania discolor, V. vincta;* **Curculionidae**–*Baris dolosa, Microlarinus* sp.; **Nitidulidae**–*Macroura* sp.

Diptera
(Flies): Asilidae–*Laxenecera* sp.; **Bombyliidae**–*Euchrimyia* sp., *Geron* sp., *Ligyra sphinx;* **Calliphoridae**–*Calliphora* sp., *Chrysomya megacephala, Rhyncomya viridaurea, Stomorhina discolor;* **Chloropidae**–*Anatrichus pygmaeus;* **Ctitidae**–*Physiphora* sp.; **Milichiidae**–*Milichiella* sp.; **Muscidae**–*Musca crassiirostris, M. domestica, M. pattoni;* **Sarcophagidae**–*Sarcophaga orchidae, Sarcophaga* sp.; **Syrphidae**–*Episyrphus balteatus, Eristalis laetus, E. quinquestratus, E. tenax, Eumerus albifronss, E. nepalensis, Ischodon scutellaris, Metasyrphus corollae, Microsyrphus confacter, Seaeva indiana, S. pyrastri, Sphaerophoria benghalensis, Syrphus* sp.

Hymenoptera
(Ants): Formicidae-*Camponotus compressus, C. sericeus, Camponotus* sp., *Crematogaster* sp., *Formica* sp., *Monomorium indicum, Oecophylla samaragdina, Paratrechina* sp., *Solenopsis geminata.*
(Wasps): Eumenidae – *Delta conoideus, D. pyriformis, Rhynchium metallicum;* **Scolidae** – *Campsomeris* sp., *Megascolia flavifrons, Scolio cruenta;* **Sphecidae** – *Bembix borrei, Bembix* sp., *Chalybion bengalense, Sphex* sp.; **Vespidae** – *Polistes habreus, P. stigma tamula, Ropalidia spatulata, Ropalidia* sp., *Vespa* sp.

(Bees): Apidae–*Apis cerana indica, A. dorsata, A. florea, A. mellifera, Bombus albopleuralis, B. asiatica, B. avinoviellus, B. bahumicus, B. biroi, B. branickii, B. ferganicus, B.flavothoracicus, B. haemorrhoidalis, B. himalayans, B. hyponourus, B. infofasciatus, B. kashmirensis, B. keriensis, B. ladakhensis, B. lemnscatus, B. lepidus, B. lucorum, B. morawitzians, B. marussinus, B. melanurus, B. novus, B. oberti, B. peronatus, B. pyrosoma, B. reimiga, B. semenovianus, B. simillimus, B. subtypicus, B. trifasciatus, B. tunicatus;* **Andrenidae**–*Andrena flavipes, A. ilerda, A. leaena, Trigona* sp.; **Anthophoridae**–*Amegilla quadrifasciatus, Amegilla* sp., *Anthophora* sp., *Centris inermis, Ceratina simillima, Melissodes* sp., *Pithitis binghami, P. smaragdula, Thyreus histrio, Xylocopa fenesterata, X. latipes, X. pubescens, X. rufescens, X. valga;* **Colletidae**–*Colletes* sp.; **Halictidae**–*Halictus* sp., *Lasioglossum* sp., *Nomia strigata, Pseudapis oxybeloides;* **Meghachilidae**–*Megachile bicolor, M. cephalotes, M. flavipes, M. lanata, M. nana, M. femorta, Osmia* sp., *Threus* sp.

Lepidoptera
(Moths): Sphingidae–*Macroglossum gyrans*
(Butterflies): Acraeidae–*Acraea violae;* **Danaidae**–*Danaus chrysippus, Euploea core, Tirumala limniace;* **Hesperiidae**–*Borbo bevani, B. cinnara, Gangara thyrsis, Pelopidas mathias;* **Lycaenidae**–*Castalius rosimon, Everes parrhasius, Euchrysops cnejus, Rapala jarbus, Spindasis vulcanus;* **Nymphalidae**–*Hypolimnas bolina, H. misippus, Junonia almana, J. hierta, J. lemonias, , J. orithyia, Phalanta phalantha;* **Papilionidae**–*Papilio polymnestor, P. polytes, , Princeps demoleus, Pachliopta hector, P. aristolochiae, Graphium agamemnon, G. doson, G. nomius;* **Pieridae**–*Anaphaeis aurota, Catopsilia crocale, C. pyranthe, C. Pomona, C. Pomona catilla* form, *Cepora nerissa, Colotis danae, C. eucharis, Eurema hecabe, Leptosia nina, Valeria valeria.*

Beetles

Beetle-pollination is characteristic of tropical environments. Beetles were the first pollinators of flowering plants and Diels (1916) was the first to hypothesise that beetle-pollination system is characteristic of the ancestral angiosperms. Diels' interpretation of the early angiospermous flowers as beetle pollinated was turned into a conviction, and the results of recent studies on pollination mechanism of extant primitive angiosperms Gottsberger (1988) corroborated Diels' view that beetle pollination is prevalent in primitive angiosperms. It is generally held that the beetles are the least specialized flower visitors and accomplish pollination in a mess-and-soil manner. They visit flowers primarily for edible rewards (pollen and nectar), mating activities, and thermogenesis. Bernhardt (2000) divided beetle-pollinated angiosperms into four overlapping modes: bilabiate, brush, chamber blossoms, and painted bowl. It is the chamber blossom flowers that abound in the basal angiosperms that have been traditionally used to illustrate beetle pollination in 'primitive angiosperms' (Thien *et al.*, 2000). But the characters may not be uniform with all the cantharophiles. Even in Magnolidae, diversity in the syndrome was noted and different beetle-flower interactions have been identified (Thien, 1980; Thien *et al.*, 2000). It may then become necessary to distinguish between different syndromes of beetle pollination as suggested by Bernhardt and Thien (1987). Guided by their own work on *Myristica insipida*, Armstrong and Irvine (1989) proposed the term 'Microcantharophily' to distinguish the same from 'typical' cantharophily. We found in India the scarabeid beetles as the primary pollinators of *Chrozophora rottleri* (Euphorbiaceae), a weed of the dry arid habitats. The Chrozophora-scarabeid beetle system does not totally agree with either of the two syndromes recognised; some features correspond with cantharophily and some with microcantharophily. Characteristics identical with cantharophily include limited flower number, aggregation of beetles, prolonged visits, mating within the flower and visitor constancy; those matching microcantharophily include small, inconspicuous, exposed flowers, offering pollen as a reward and the small beetles foraging on pollen. Beetle pollination in *Chrozophora* thus represents a different system. *Euphorbia geniculata* received beetles of the families Coccinellidae, Nitidulidae, Bruchidae, Curculinidae (Reddi and Subba Reddi, 1984). These two plant species belong to Euphorbiaceae, a phylogenetically advanced family.

Flies

Flies are also the pollinators of primitive angiosperms (Thien *et al.*, 2000). Of the three suborders of Diptera, the Nematocera are considered to be the most primitive, and their relations with flowers is rather poorly known (Proctor and Yeo, 1972). However, there are certain efforts showing the importance of Nematocera in flower pollination — fungus gnats and flowers; mosquitoes and *Plantanthera* (= *Habenaria*) orchids; various nematocerans and Araceae; and Ceratopogonidae and Cocoa (see Kevan and Baker, 1999); gall midges and *Siparuna* (Feil, 1992). The second suborder Brachycera consists of a wide diversity of flower visitors. Some Brachycera possess short mouthparts and some long mouthparts, the former making visits to many species bearing shallow flowers with easily accessible nectar, and the latter visiting deep flowers with tubular corolla. The Bombylidae have long sucking mouthparts capable of handling tubular flowers, and are prominent among the flower-visiting Dipterans. The third suborder Cyclorrhapha is divided into two series, Aschiza and Schizophora. In the Aschiza, Syrphidae (hover or flower flies) represents the most important family of flower visitors. These flies feed on nectar and pollen. Little is known of other features. The series Shizophora is split into two sections, Acalypterae and Calypterae. In the Acalypterae, the family Conopidae is noteworthy for its flower-visiting members, but their possible role in pollination is not well understood. The Calypterae consists of important flower visitors; particularly the huge family Muscidae and Calliphoridae are known anthophiles (Proctor and Yeo, 1972). Kevan and Baker (1999) expressed that the role of Calypterae in pollination is discounted despite their abundance, frequency of visits and effective pollen transport. In our study the members of Calypterae are proved to be the exclusive pollinators of *Sterculia foetida*, and also the major pollinators of *Euphorbia antiquorum*, and *Zizyphus mauritiana*. The ideal fly-flowers are expected to produce nectar in small aliquots, and as the nectar is freely exposed post secretary evaporation causes nectar to be concentrated (Corbet, 1978). In both *E. antiquorum* and *S. foetida*, nectar is secreted in traces and as the blossoms are shallow and exposed, the moisture in nectar may get evaporated and the nectar may get concentrated, making the blossoms more attractive to flies. Faegri and Pijl (1979) treated flies because of their limited flight range and feeding habits as unspecialised and unreliable as pollinators. The plant species serviced by such flies may possess adaptations to achieve outcrossing through mechanisms of protogyny, protandry, dioecism (Ashton, 1969), and temporal dioecism or duodichogamy (Cruden, 1988). *Sterculia foetida*, though morphologically andromonoecious, is functionally dioecious due to incompatibility to geitono-pollen and *Euphorbia antiquorum* displays temporal dioecism. In both these plant species, *Chrysomya megacephala* proved to be abundant and effective in promoting cross-pollination and the attendant genetic variability. It may be of interest to note that *Chrysomya* is the proven pollinator of the Southeast Asian parasitic flowering plant *Rafflesia* known for its largest flowers (Beaman *et al.*, 1988). Cashewnut crop is pollinator limited, which can be remedied by managing the flies by growing *Euphorbia antiquorum* as a hedge; *E. antiquorum* comes to bloom just prior to cashewnut and maintains fly population which can move to cashew in bloom (Reddi, 1987).

Ants

Ants are extremely abundant in the tropical habitats. But adaptations to pollination by ants are considered to be rare and it is generally held that they are very seldom pollen-vectors (Beattie, 1982). The rarity of ant-pollination is attributed in the first instance to the limited foraging areas and the lack of wings on the worker ant, therefore they cannot effect gene flow between and among the plants they service (Feagri and Pijl, 1979; Armstrong, 1979). The other reason is based on the finding that secretions on the surface of the ants are deleterious to pollen. The pollen exposed to the ant integument for brief periods had reduced viability and germination relative to control affected by the "myrmicacin" secretions of the ant's metapleural gland. Beattie *et al.* (1984, 1985) suggest that these secretions of myrmicacin incidentally cause disruption of pollen function militating against the evolution of ant-pollination. However, there is evidence that ant-pollination may be extremely common in a few kinds of habitat. Taking the suggestion of Bentley (1983) that nectar may play an important role not only in insect nutrition but also in water balance as well, Rico-Gray (1989) proposed that in the tropics particularly in the dry season, floral nectar may be most attractive both as an energy and water source to the ants. It is therefore expected that ants may prove to be important pollinators of the plants inhabiting dry tropical areas and flowering during the dry season. Hagerup (1943) working in Southern Sahara, and Hickman (1974) in Western Cascades of Oregon predicted the occurrence of ant-pollinator interactions on hot and dry habitats where the frequency of activity of ants would be normally high. Pollination of the orchid *Loperella fimbriata* by pseudocopulation by winged males of the ant *Myrmecia urens* studied by Peakall *et al.* (1987) weakens any argument against the evolution of ant pollination. These authors suggest that ant-pollinators do not possess metapleural glands or the plants' pollen must possess resistant characters. Metapleural glands are absent in all castes of the genus *Camponotus* (Holldobler and Engel-Siegal, 1984). Under Indian conditions, *Euphorbia dracunculoides*, a weedy species of the dry arid lands, has all the adaptations for pollination by the ant *Camponotus* species. The cyathia are small (2.5–3.0 mm long; 1.75–2.00 mm wide), possess no odour, and secrete nectar in traces, thus unlikely to attract pollinators of high-energy demand. The plants are small (20–40 cm high) and occur in extremely dense populations, almost monotypic, thereby reducing ant-foraging distances. When the populations are thin, there is excessive branching; thus interplant distances are minimised and pollen movement between plants by the low energy requiring crawling ant is assured. The worker ants are extremely agile and intensively forage at the cyathia. Nectar foraging by the winged males largely promoted xenogamy. In another species *Euphorbia geniculata* ants numerically preponderated and are the only pollinators when the plants occur in fields cropped to *Sorghum*, *Pennisetum* and Maize. Fruit or seed set or fecundity under ant pollination and others did not show any difference, thereby showing a definitive role of ants in pollination (Reddi and Subba Reddi, 1984).

Wasps

Despite the sensational, curious and fascinating mode of pollination in certain orchid species through pseudocopulation by certain wasps, and the one-to-one relationship of the Fig species and the Agaonid wasps (*Blastophaga*), the wasps in general are treated as unreliable and unsteady pollinators because a great majority of the true wasps are primarily animals of prey and feed their larvae with insect's larvae (Proctor and Yeo, 1972; Feagri and Pijl, 1979). Quite often, the wasps visit the flowers for nectar to meet their carbohydrate requirement. At least some wasp species feed their larvae with nectar, and this sugary fluid is essential to the diet of *Vespula* larvae (Brian and Brian, 1952). In some higher wasps like *Polistes*, mouth parts are modified into tubular structures as against the flat tongue of many wasp species and through these tubular structures nectar can be sucked up and the species possessing such mouth parts are quite capable of using blossoms and pollinating them (Feagri and Pijl, 1979). Our studies proved the role of wasps in the pollination of such plant species as *Premna latifolia*, *Santalum album*, *Zizyphus mauritiana*, *Z. oenoplia*, *Vitex negundo*, etc. Though these floral species are promiscuous receiving visits from many insect species, they are rather more attractive to wasps and are persistently and abundantly visited by diverse wasps. Some of the wasps avoid the legitimate way of foraging and resort to nectar robbing through making the holes in the corolla. *Rhynchium metallicum* and *Ropalidia* species adopt this behaviour to reach the nectar in flowers of *Vitex negundo*. Generally nectar robbing is restricted to circumstances that limit tongue lengths to reach nectar. Here corolla entrance of *V. negundo* is wide enough to permit the two wasps to push their heads into the flowers and suck nectar. Yet these wasps resorted to puncturing the corolla and take nectar through perforations. Other foraging wasps like *Bembix* also use these perforations for nectar uptake though they do not create perforations. Other tropical studies also reveal a larger role of wasps in pollination than is normally believed (Baker *et al.*, 1983).

Bees

The bees, whether social or solitary, totally depend on flowers and to quote Kevan (1995), "the flowers are the mainstay of bees' life". The bees require enormous quantity of pollen and nectar for the nutrition of themselves and for their brood. Hence they make numerous visits to flower and intensively forage for the said two floral resources. Morphologically and behaviourally, they are well suited to collect pollen and nectar. On this count they are lauded as the most important pollinators of a large proportion of angiosperms. The bees are diurnal in their foraging activity, but can also work in moonlit nights (Sihag, 1984; Rao *et al.*, 2001). The rockbee *Apis dorsata* is the sole visitor and pollinator of the night blooming *Pterocarpus santalinus* (Fabaceae) in its natural habitat. This tree species displays mass flowering in few episodes and the flowers open at around midnight and are foraged during moonlit hours. The carpenter bee *Xylocopa rufescens* is the sole pollinator of night blooming *Antirrhinum majus* (Scrophulariaceae) (Bhaskar and Gopinath, 1975).

Most studies in India are concentrated on bee pollination of agricultural crops with a view to enhancing crop productivity. Bee pollination of these crops including the self-pollinated crops was observed to greatly enhance productivity of crops (Deodikar and Suryanarayana, 1977; Mishra, 1989; Abrol, 1993). The domesticated honey bee *Apis cerana indica* and the imported European *Apis mellifera* were largely tested for their efficacy in pollinating the crop species and improving their yields. In the event of large acreage under cultivation of any crop as is the case with sunflower, management of honey bees proved highly rewarding. Non-agricultural plant species are rather totally neglected, and only recently have there been few studies focusing on natural vegetation as a source of floral resource to the honey bees and this has a bearing on nectar harvesting and honey production. Describing the insects and flowers, Burkill (1906) observed the Xylocopas as the most important of flower visiting insects in the plains of the country and the work is taken up by *Bombus* in the hills particularly the Himalayas. The Sun hemp is largely fertilised by the Xylocopas (carpenter bees), and possibly the Indian pulses are also effectively pollinated. Studies at Andhra University demonstrated an extremely important role for the carpenter bees in carrying out pollination in Cassias, Lamiaceae, Acanthaceae, and Verbenaceae. Most of the floral species handled by these carpenter bees are zygomorphic with a mechanism to deposit the pollen in a specific location (notum) on the pollinator from which it is least likely to be removed during body grooming. This operation is successful in the nototribic flowers, and the bilateral symmetry of the body of Xylocopas is quite suited for such floral structures. The role of Xylocopas, and possibly of other anthophorid bees, is of paramount importance in the pollination of floral species with poricidal anthers. The bees sonicate the anthers using trains of audible buzzes generated by the thoracic muscles. This mechanism results in the buzz-pollination of nearly 20,000 species of plants the world over (Buchmann, 1983). The Cassias are well suited for the buzz pollination by different species of *Xylocopa* (Gottsberger and Gottsberger, 1988). The nectarless *Cochlospermum religiosum* with big yellow flowers possessing numerous anthers is cross-pollinated exclusively by *Xylocopa latipes* and *X. pubescens* (Bhaskara Rao and Subba Reddi, 1994). The honey bees are unable to collect the pollen of poricidal anthers by the usual method of brushing and brooming. Thus there appears to be a partitioning of resources between honey bees and solitary bees, particularly the Xylocopas. Another solitary bee capable of buzz collecting of pollen in such plant species as *Solanum* and *Lycopersicum* is the digger bee *Amegilla*. The stingless bee *Trigona* is also known for its buzz collecting of pollen and is a proven and substantial contributor to pollination of atleast nine crop species including *Mangifera indica* (Mango) and *Cocos nucifera* (Coconut) (Heard, 1999). Several solitary bee species are implicated in the pollination of several agricultural crops, and these include *Amegilla, Andrena, Megachillae, Halictus, Lasioglossum, Colletes,* etc. The small stingless bees (Meliponinae), honey bees (Apinae) and halictids are often described and discounted as cross-pollinators because of their opportunistic behaviour, low energy demand with short flight ranges, but they cannot be underestimated to serve as cross-pollinators (Renner and Fiel, 1993). These bees are known to respond dramatically to spatial and temporal changes in floral resource and move 'enmasse' with the pollen adhered to them to other conspecific plants in flower (Bawa, 1980). This kind of movement has been demonstrated in the mass blooming *Terminalia tomentosa* where nectar resorption occurs simultaneously in all open flowers (Atluri *et al.*, 2000). These small bees are the visitors of many obligately outcrossing plant species that also are dioecious in the tropics (Roubik, 1989).

Bees belong to the super family Apoidea, with over 20,000 species in nine families worldwide (Michener, 1974). In India the species so far identified are in 47 genera distributed in seven families as Colletidae, Halictidae, Andrenidae, Melittidae, Megachilidae, Anthophoridae and Apidae (Batra, 1977). While commenting on the buzzy diversity of bees in India, Batra (1993) expressed that the flower relationships of the diverse Apoidae have been neglected. It is hoped that there surely remain numerous undescribed species functioning as pollinators of diverse natural vegetation.

Moths

The ethodynamic pollination of *Yucca* by the moths of *Tegiticula* (*Pronuba*) is well known. Recent studies established that hawkmoth pollination may be more common in tropical habitats (Opler, 1983), but not confined to the tropics (Grant, 1983). Hawkmoths may be important pollinators in orchid species (Johnson, 1995), *Capparis* species (Eisikowitch *et al.*, 1986) and *Datura meteloides* (Johnson, 1995). The moths are crepuscular or nocturnal, but some are diurnal; accordingly the former use nocturnal flowers and the later forage on the diurnal, and also nocturnal flowers that stretch into the following day. Hawkmoths are strong and rapid fliers, hover in front of their flowers while taking the nectar hidden in long narrow corolla tubes. Floral and nectar features of the dioecious *Xeromphis spinosa* in India complies with these characters and thus appears to be well tailored for manipulation by the sphingid hawkmoth, *Macroglossum gyrans*, which is diurnal and crepuscular in activity. There is a close match between temporal activity of hawkmoth and the opening of flowers, both events being more pronounced at dawn and dusk. The flowers available at these times contain more nectar while others secrete traces. This nectar secretion pattern reinforces the relation between *X. spinosa* and its hawkmoth pollinator *Macroglossum gyrans* (Raju *et al.*, 1997). A similar activity pattern is also reported on *Martynia annua* (Bhaskara Rao and Subba Reddi, 1994) and *Duranta repens* (Byragi Reddi and Subba Reddi, 1996). Other plant species with adaptations for moth-pollination include *Carissa* species, *Holarhena antidysentrica*, *Morinda tinctora*, *Nyctanthes* species and Oleaceae members (Bhatnagar,

1989). The diurnal hawkmoth *M. gyrans* has been found to forage on *Tecoma* species, *Lantana camara* and *Jasminum angustifolium*, etc. *Sansevieria* of Liliaceae which effloresces in the month of July (August), has its flowers opening between 1700–1800 h; it is. foraged and pollinated within another half an hour exclusively by the diurnal hawkmoth *M. gyrans*. The noctuid moths pollinate *Barringtonia acutangula*. Interestingly the corolla is cup-shaped with short lobes, and the nectar is exposed, and the stamens are long with red colour.

Butterflies

Most butterflies visit flowers for nectar during daytime. Nectar provides energy and amino acids for nutrition, increases butterfly longevity and egg production and also improve egg maturation (Watanabe, 1992 and the references therein). While Wiklund *et al.* (1979) argue that "butterflies as a group may have evolved to occupy a parasitic mode of life as adults feeding on the nectar of flowers without pollinating them," several authors show that butterflies do mediate pollen transfer as in *Dianthus deltoides* and *Viscaria vulgaris* (Jennerstern, 1984), *Phlox* spp. (Levin and Berube, 1972), *Caesalpinia pulcherrima* (Cruden and Hermann-Parker, 1979), *Platanthera* spp. (Smith and Snow, 1976) and *Cnidoscolus urens* (Bawa *et al.*, 1983). In South Indian plants, important butterfly pollinators include *Danaus chrysippus, Catopsilia crocale, C. pyranthe, Precis lemonias, Papilio polytes*, etc. (Balasubramanian, 1992). Whether a butterfly visiting a floral species participates in the pollination play depends on floral architecture and intrafloral behaviour of the butterfly (Subba Reddi and Meera Bai, 1986). The Papilionids hover directly in front of the flower while probing for nectar whereas the other butterflies perch on the flower/inflorescence and probe for nectar. Further, depending on the floral architecture, the position of essential organs relative to other floral parts, the foraging position of butterflies relative to anthers and stigma vary from species to species. Consequently, contact of essential organs of a flower cannot be established at a fixed point on the butterfly's body. It may be a proboscis, head, antennae, legs and wings. Pollen deposition on proboscis and head is treated as an important trait of psychophily (Kevan, 1997). Of the 30 Indian butterfly species examined in this context, all the species received pollen on proboscis (and head) in a majority of the floral species they probed for nectar, thereby indicating their potential as pollen-vectors (Table 8). The time spent per flower, and the number of flowers visited per minute are indicative of the foraging speed, and in this parameter the Papilionids excel other butterfly families (Table 9). The Pierids

Table 8: Different Body Parts of Butterflies that Pick-up Pollen While Nectaring.

Butterfly species	*No. of plants visited*	*Proboscis*	*Head*	*Antennae*	*Legs*	*Wings*
ACRAEIDAE						
Acraea terpsicore	12	12	6	6	6	8
DANAIDAE						
Danaus chrysippus	30	26	18	6	14	5
D. limniace	7	7	5	3	5	0
Euploea core	16	14	5	4	7	0
HESPERIIDAE						
Borbo cinnara	22	20	11	2	8	0
Pelopidas mathias	11	10	6	2	4	0
LYCAENIDAE						
Castalius rosimon	13	10	6	3	8	5
Euchrysops cnejus	7	4	3	1	4	0
Jamides celeno	3	2	1	0	2	0
NYMPHALIDAE						
Euthalia garuda	3	3	3	0	3	3
Hypolimnas bolina	5	5	3	3	4	3
H. misippus	7	7	3	2	4	0
Phalanta phalantha	16	15	7	4	7	6
Junonia almana	6	6	3	2	3	0
J. hierta	11	11	5	5	5	5
J. lemonias	15	14	8	5	10	0
PAPILIONIDAE						
Graphium agamemnon	13	12	4	1	2	3
Pachliopta aristolochiae	30	24	10	3	5	3
P. hector	19	16	3	0	3	0
Papilio polytes	23	21	5	3	8	8
Princeps demoleus	15	14	4	0	3	6
PIERIDAE						
Anaphaeis aurota	9	6	3	2	2	0
Catopsilia crocale	11	11	4	0	2	0
C. pyranthe	15	11	14	6	9	6
Cepora nerissa	11	11	2	1	4	3
Colotis eucharis	11	10	5	1	2	2
C. danae	9	8	4	1	1	1
Eurema hecabe	18	16	6	3	6	0

Table 9: Time Spent (seconds) per Flower and Number of Visits per Minute by Each Butterfly Family.

Butterfly family	*Time spent/flower*	*No. of visits/minute*
Papilionidae	2.4	19
Pieridae	6.3	6
Nymphalidae	9.7	5
Danaidae	8.2	4
Hesperiidae	13.1	3
Lycaenidae	13.8	3

vied for the second position. The Lycaenids stood in the last position. Certain floral species possess rather elongated staminal filaments and style (sometimes gynandrophore) and orient their flowers horizontally. Such floral configurations occur in *Ceasalpinia pulcherrima, Clerodendrum infortunatum, C. inerme, Delonix regia* and *Cadaba fruiticosa*. The former four species are mostly visited by Papilionids, and the latter by the Pierids *Colotis* and *Anaphaeis* (Subba Reddi and Meera Bai, 1986). In *C. infortunatum* and *Delonix regia* the big bodied *Papilio polymnestor, Pachliopta aristolochiae, P. hector* and *Papilio polytes* perform pterygotribic pollination (Byragi Reddi and Subba Reddi, 1995). In *C. fruiticosa* and *C. pulcherrima* nectar is placed in a special container (nectarial tube), and the spatial separation of this and essential organs is of positive value for the correct use and placing of proboscis (Cruden and Hermann-Parker, 1979). These four species are tailored for manipulation and pollination exclusively by the butterflies, of course not all butterfly families can handle these flowers and effectuate pollination. May be in certain localities because of infrequent visits, butterfly pollination is not quantitatively important (Courtney *et al.*, 1982). However, the drifting behaviour of butterflies to visit a few flowers on a plant and fly away contributes to xenogamous pollinations vis-a-vis gene flow distances as demonstrated in *Senecio* (Schmitt, 1980). It is commonly held that butterflies prefer nectar concentrations in the range of 20–25 % sucrose, the nectars above this range becoming increasingly unprofitable (Pyke and Waser, 1981). Recent studies have shown, however, that butterflies actually are able to feed intensively on nectars having 30–65% sucrose and that optimum nectar concentration is approximately 40% (Pivnick and Mc Neil, 1985; May, 1985; 1988). The rate of extracting nectar is related to body size (Herrera, 1989). Butterflies with a larger body size can extract nectar at a faster rate than small body size butterflies, and this behaviour is reflected in the flower handling time. The Papilionid butterflies normally have a larger body size, able to have fast nectar extracting and thus to have greater foraging speed, in that they cover a large number of flowers in a foraging trip.

Birds

As early as 1924, Porsch recognised eight most important families of flower-visiting birds of the world; these are listed in Proctor and Yeo (1972) giving their geographical distribution. A review by Subramanya and Radhamani (1993) lists 58 flower-visiting Indian bird species distributed in 16 different families (Table 10). Out of the 8 families of

Table 10: Bird Pollinators in India

Bird family	*Bird species*
Psittacidae	*Psittacula krameri, Loriculus vernalis*
Currulidae	*Eudynamys scolopacea*
Picidae	*Dinopium benghalense*
Laniidae	*Lanius schach*
Oxiolidae	*Oriolus oriolus, O. xanthornus*
Dicruridae	*Dicrurus adsimilis, D. lecophaeus, D. caerulescens, D. hottentotus, D. paredisens*
Sturnidae	*Sturnus malabaricus, S. pagodarum, S. roseus, Acridotheres tristis, A. fuscus, Gracula religiosa*
Corridae	*Corvus splendens, C. macrorhynchos, Denohocitta vagabunda*
Irinidae	*Aegithina tiphia, Chloropsis aurifrons, C. cochinchinensis*
Pycnonotidae	*Pycnonotus jocosus, P. cafer, P. luteolus, Hypsipetes madagascariensis*
Muscicapidae	*Xiphirynchus supercillaris, Turdoides striatus, T. affinis, Dumetica hyperythraalbogularis, Chrysomma sinensis, Muscicapa tickelliae, Rhipidura albicollis, Terpsiphone paradise, Monarcha azurea, Pinia subflavu, P. socialis, Orthotemus sutorius, Copsychus saularis, Turdus merula migropileus, Dicaeum erythrorhynchos, D. concolor*
Nectarinidae	*Nectarinia zeylonica, N. minima, N. lotenia, N. asiatica, Aethopyga siparaja, Arachnothera longirostris*
Zosteropidae	*Zosterops palpebrosa*
Ploceidae	*Passer domesticus, P. xanthocollis, Ploceus philippinus, Lonchura malabanica, L. striata*
Fringillidae	*Carpodacus erythrinus*

the flower-visiting birds of the world, four families (Nectariniidae, Zosteropidae, Dicaeidae, and Psittacidae) are represented in India, and the other families (Trochilidae, Coerebidae, Meliphagidae and Drepanididae) are absent. The main flower-visiting birds in Europe are Warblers of the sub family Sylvinaea of the family Muscicapidae, which also occur in India. Birds of similar status are Hummingbirds in the Neotropics, Honeyeaters, Lories and Lorikeets in the Australia-Papuan region, and sunbirds in Africa. The 58 Indian flower-bird species are reported to visit 93 flowering plant species (Subramanya and Radhamani, 1993), and to this list may be added *Brugiera* species and *Acanthus ilicifolius* of mangrove forest, and *Anisomeles indica* and *A. malabarica* occurring in semi rocky areas. *Brugiera* with explosive opening of flowers is visited and pollinated by white-eye, and the other three plant species by sunbirds. Out of the 93 plant species, the birds are named for only 78 species. When the number of plant species visited by each bird species are analysed, the two sunbird species — *Nectarinia asiatica*, and *N. zeylonica* came out as the versatile flower-visiting bird species foraging at 73–74% of the 78 flowering plant species. However, the role of these birds as pollen vectors of flowering plants they visit is yet to be documented in detail. Only one study obtained details of the effective link between birds and mistletoes (Loranthaceae) (Davidar, 1985). In Loranthaceae all dioecious species are entomophilous, whereas all hermaphroditic species are bird-pollinated (Kuijt, 1969). The mistletoe flowers normally require an external pressure to open, and such pressure comes from bird visitors, which ultimately help the flowers in pollination. But spontaneous flower opening occurs in *Helixanthera intermedia* and *Dendropthe memecybifolia*. She showed that four species of birds — flower pecker (*Dicaeum concolor*), white-eye (*Zosterops palpebrosa*), small sunbird (*Nectarinia minima*) and purple sunbird (*Nectarinia asiatica*) as the effective pollinators of seven species of mistletoes, with close knit evolution between specific bird and mistletoe species in Southern India. Sunbirds maintain flower constancy. *Spathodea companulata* and the entire genus *Erythrina* is ornithophilous and the flowers are adapted to pollination by different passerine birds. Wesley (1987) reported 12 species of birds visiting the blossoms of coral tree *Erythrina indica*, of which *Sturnus payodarum* (black-headed myna) was the commonest. The author found much pollen wastage due to the damage caused to floral parts and poor pollination by the birds. In a detailed study on the pollination of *Helicteres isora*, often cited as the bird pollinated species, Atluri *et al.* (2000) showed flower colour changes, from bluish-grey (day 1) to red (day 2) and deep red (day 3). The tubular flowers with rather a wide mouth are probed by insects mostly for pollen and birds for nectar. The bird species *Turdoises affinis* and *Acridotheres tristis* probe the blue-grey flowers occasionally, red flowers very frequently and deep red flowers frequently. While nectaring the two bird species contacted the anthers and stigma with their bill and getting powdered with pollen on bill/forehead invariably effected pollination. Considering their probing behavior and the manner of effecting pollen transfer, these two bird species are treated as appropriate pollinators. Another bird species *Pycnonotus cafer* was found to cut the anterior part of the staminal column and gynophore with anthers and stigma with its beak prior to probing the flower for nectar, hence its visits are detrimental to plant's reproduction. These studies suggest that all bird visits may not be of help in pollen transfer, and a closer study is needed to establish the adaptiveness for bird-pollination. Westerkamp (1990) discussed the hovering and perching methods of flower exploitation by birds and distinguished the hovering behaviour, believed to be confined to Trochilidae, into

"Schwirr flug" (whirling flight of trochilids and "triller flug" (trilling and fluttering flight) of other birds. Kannan (1978) studied the bill and tongue lengths of flower birds, and divided them into two natural groups: the non-specialised nectar feeders with licking type of bill and tongue, and specialised nectar feeders with suctorial mouthparts. *Chloropis aurifrons, C. cochin chinensis, Dicaum erythrorhynchos,* five species of sunbirds and white-eye are the specialised nectar feeders. The Dicaeidae, Nectarinidae and Zosteropidae are phylogenetically related ones and are aptly described as the "Asiatic nectar feeding assemblage". Despite the reported association of several bird species with several floral species, an obligate association of birds and floral species appears to be rather uncommon. Since the diet of flower birds include not only nectar as an energy source but also insects to satisfy their protein requirement (Ali, 1988), obligate reliance on birds species for pollination may not be an ideal fitting strategy. Most of the floral species identified as bird flowers (*Moringa oleifera, Bauhinia purpurea, Peltophorum pterocarpum, Cassia* species, *Calotropis gigantea, Tecoma stans, Adhatoda zeylanica, Acanthus ilicifolius, Anisomeles* species, *Gmelina arborea* and *Cochlospermum religiosum*) have been proved to be pollinated by carpenter bees. Such observations have led to a hypothesis that the evolution of bird-pollination (ornithophily) followed that of entomophily (Grant, 1994). This proposition is well substantiated by flower-pollination data of Western North America, and also by the Indian data. Nevertheless the associations: white-eye — *Brugiera*, and Asiatic nectar feeding assemblage of birds — mistletoes, may be taken as a mutualistic obligate association.

Bats

Bats are tropical in distribution and bat-pollination is typically a tropical phenomenon (Vogel, 1958). Bats constitute the systematic order Chiroptera, with two sub-orders as Megachiroptera containing the single family Pteropidae (flying-foxes) primarily adapted to frugivory, and Microchiroptera, which are largely insectivorous. Both frugivorous and insectivorous bats visit flowers for pollen and nectar. Accordingly, the flowers adapted to bat-pollination supply abundant pollen and, nectar which is rather thin. Further, flower opening is tuned to the nocturnal activity of bats. Because of the problems in studying this nocturnal phenomenon, detailed studies of the supposedly bat pollinated floral species are almost lacking. The only earlier study was of Mc Cann with the sausage tree *Kigelia pinnata* (Subramanya and Radhamani, 1993), and recently Subba Reddi *et al.* (1989) made a detailed study on *Ceiba pentandra* indicating the close correspondence between anthesis, nectar production, and bat activity. The blossoms are of shaving brush type opening at dusk from 1830 h and remaining open until the next day up to 1200 noon. Nectar production starts at flower opening and ceases by 0200 h and the bat activity closely corresponds with the nectar presentation pattern. The large fruit bat *Pteropus giganteus* which has a tendency to roost in large trees, was the first to visit flowers clustered near the tips of branches, lap nectar and eat pollen. The small bats *Cynopteris sphinx giganteus* and *Taphozous longimanus* using thatched houses for roosting made visits during the same period. The bats first landed on branches, crawled on them to reach the flowers. While foraging, the snout and abdomen of the bat brushed against the anthers and stigma, the pollen thus received was transferred to the stigmas of other flowers during the successive foraging visits. The large bats made interplant visits, promoting xenogamous pollinations. Accessibility to modern techniques of electronic flash photography, image intensifying, and night viewing devices is required for extensive and intensive studies of such floral species as *Careya arborea, Sonneratia apetala* etc., which appear to have adaptations for bat-pollination.

Conclusion

The functional sexual and pollination systems including the type of pollinators of diverse hermaphrodite, andromonoecious and monoecious plant species (some of which are phylogenetically related) are analysed. The angiospermic flora possess a wide range of selfing as well as outcrossing mechanisms. The occurrence of such diversified sexual systems in the hermaphroditic species itself leads to the evolution of 'balanced sexual system', which represents a compromise determined by the evolutionary history and evolutionary potential of the population as well as the requirements of the present environment, thus allowing permanent survival of the population. Then the most ideal sexual system can be the one with a provision for a mixing of crossing and selfing permitting species survival and maintenance of genetic diversity. Such a 'mixer of crossing and selfing can keep the evolutionary pot boiling nearly as well as outcrossing alone'. Naturally most angiosperms might have developed a mixed or flexible sexual system enabling them to cope up with the changing environment. The present analysis also revealed that pollen vectors or pollinators of a vast number of naturally occurring plant species in different habitats of the climatically and ecologically varied land of India remain virtually unexplored and thus there is ample scope even for the amateurs to produce rewarding results. It is increasingly realized that pollinators play an important role in the maintenance of natural vegetation through mediating pollinators, the vital ecological process both for natural and agricultural sustainability of the world. Much is required in the complete identification of the diverse biotic pollinators, and this requires a coordination between botanists and zoologists/entomologists of the country. Among the diversified biotic pollinators including thrips, beetles, flies, ants, wasps, bees, moths, butterflies, birds and bats, the large pollinators with high metabolic requirement ('high energy pollinators') tend to fly longer distances than small pollinators with low metabolic requirement ('low energy pollinators'), with the consequent effects on the genetic neighbourhood size. The ants possess crawling habit and are thus low energy pollinators. The large sized pollinators like carpenter bees, papilionid butterflies, hawkmoths, birds and bats are the high energy pollinators. These high energy pollinators are also known for their 'traplining' behaviour, flying and promoting gene flow between distantly placed conspecific plants. These wild pollinators can be the keystone species. These are being exposed to the impact of indiscriminate application of pesticide, destruction of natural areas, promotion of monocultures and clean cultivation causing the elimination of pollinator nesting sites and food resources. It is to be realised that conserving the wild pollinators will help conserve and maintain genetic diversity in the plants. It is suggested to have intensive and extensive studies of the sexual and pollinating systems with accurate identification of the pollinators in different biozones of the country for a proper understanding of how Indian angiospemic flora accomplishes and maintains its rich diversity. In the present day interests of conserving and improving plant diversity, such knowledge is urgently warranted for the effective *in situ* and *ex situ* conservation of the several endemic and endangered components of Indian angiospermic flora.

REFERENCES

Abrol, D.P. 1993. Insect pollination and crop production in Jammu and Kashmir. *Curr. Sci.*, **65:** 265–269.

Ali, S. 1988. The Book of Indian Birds. Bombay Natural History Society.

Allard, R.W. 1965. Genetic systems associated with colonising ability in predominantly self-pollinated species, pp. 50–76. In: The genetics of colonizing species. Eds. Baker, H.G. and Stebbins, G.L. Academic Press.

Ananthakrishnan, T.N. 1982. Thrips and pollination biology. *Curr. Sci.*, **51:** 168–172.

Ananthakrishnan, T.N. 1993. The role of thrips in pollination. *Curr. Sci.*, **65:** 262–264.

Armstrong, J.A. 1979. Biotic pollination mechanisms in the Australian flora — a review. *N. Z. J. Bot.*, **17**: 467–508.

Armstrong, J.A. and Irvine, A.K. 1989. Flowering, sex ratios, pollen-ovule ratios, fruit set and reproductive effort of a dioecious tree, *Myristica incipida* (Myristicaceae) in two different rain forests communities. *Amer. J. Bot.*, **76**: 74–85.

Ashton, P.S. 1969. Speciation among tropical forest trees: some deductions in the light of recent evidence. *Biol. J. Linn. Soc.*, **1**: 155–196.

Atluri, J.B., Rama Devi, K. and Subba Reddi, C. 1995. Pollen flow in *Zizyphus oenoplia* Mill (Rhamanceae). *J. Palynol.*, **31**: 261–274.

Atluri, J.B., Rao, S.P. and Subba Reddi, C. 2000a. Pollination ecology of *Helicteres isora* Linn. (Sterculiaceae). *Curr. Sci.*, **78**: 713–718.

Atluri, J.B., Rao, S.P. and Subba Reddi, C. 2000b. Self-incompatibility, nectar resorption and cross-pollination in the Indian laurel *Terminalia tomentosa* (Combretaceae). *Int. J. Ecol. Environ. Sci.*, **26**: 27–36.

Atluri, J.B., Venkata Ramana, S.P. and Subba Reddi, C. 2003. Explosive pollen release, wind-pollination and mixed mating in the tropical tree *Shorea robusta* Gaertn. f. (Dipterocarpaceae). *Curr. Sci.* (accepted).

Bahadur, B. 1968. Heterostyly in Rubiaceae: A review. J. Osmania Univ. (Sci.) (Golden Jubilee Special Volume), **4**: 207–238.

Baker, H.G. 1984. Some functions of dioecy in seed plants. *Amer. Nat.*, **124**: 149–158.

Baker, H.G. 1985. Trends in pollination biology. *Aliso* **11**: 213–229.

Baker, H.G. and Stebbins, G.L. (Eds.), 1965. The genetics of colonising species. Academic Press.

Baker, H.G., Bawa, K.S., Frankie, G.W. and Opler, P.A. 1983. Reproductive biology of plants in tropical forests, pp. 183–215. In: Ecosystems of the World-14A. Tropical rain forest ecosystems: structure and function, Ed. Golley, F.B. Elsevier Scientific Publishing Company.

Balasubramanian, M.V. 1992. Notes on butterfly pollination of some South Indian plants. *Proc. Natl. Acad. Sci. (India)*, **62B**: 561–585.

Bateman, A.J. 1956. Cryptic self-incompatibility in the wallflower: *Cheiranthus cheiri L. Heridity*, **10**: 257–261.

Batra, S.W.T. 1977. Bees of India (Apoidea), their behaviour, management and a key to the genera. *Oriental Insects*, **11**: 289–324.

Batra, S.W.T. 1993. India's buzzy biodiversity of bees. *Curr. Sci.*, **65**: 277–280.

Bawa, K.S. 1974. Breeding systems of tree species of a low land tropical *community. Evol.*, **28**: 85–92.

Bawa, K.S. 1980. Evolution of dioecy in flowering plants. *Annu. Rev. Ecol. Syst.*, **11**: 15–39.

Bawa, K.S. 1994. Pollinators of tropical dioecious angiosperms: A reassessment? No, not yet *Amer. J. Bot.*, **81**: 456–460.

Bawa, K.S. and Beach, J.H. 1981. Evolution of sexual system in tropical plants. *Ann. Mo. Bot. Gard.*, **68**: 254–274.

Bawa, K.S., Webb, C.J. and Tuttle, A.F. 1983. The adaptive significance of monoecism in *Cnidoscolus urens* (Euphorbiaceae) *Bot. J. Linn. Soc.*, **85**: 213–223.

Bawa, K.S., Webb, C.J. and Tuttle, A.F. 1983. The adaptive significance of monoecism in *Cnidoscolus urens* (Euphorbiaceae) *Bot. J. Linn. Soc.*, **85**: 213–223.

Bawa, K.S., Perry, D.R., Beach, J.H. 1985. Reproductive biology of tropical lowland rain forest trees I. Sexual systems and incompatibility mechanisms. *Amer. J. Bot.*, **72(3)**: 341–345.

Beaman, R.S., Decker, P.J. and Beaman, J.H. 1988. Pollination of *Rafflesia* (Rafflesiaceae). *Amer. J. Bot.*, **75**: 1148–1162.

Beattie, A.J. 1982. Ants and gene dispersal in flowering plants, pp. 1–8. In: Pollination and Evolution. Eds. Armstrong, J.A., Powell, J.M. and Richards, A.J. Royal Botanic Gardens.

Beattie, A.J., Turnbul, C., Knox, R.B. and Williams, E.G. 1984. Ant inhibition of pollen function: A possible reason why ant pollination is rare. *Amer. J. Bot.*, **71:** 421–426.

Beattie, A.J., Turnbul, C., Hough, T., Jobson, S. and Knox, R.B. 1985. The vulnerability of pollen and fungal spores to ant secretions: evidence and some evolutionary implications. *Amer. J. Bot.*, **72:** 606–614.

Bentley, B.L. 1983. Nectaries in agriculture, with an emphasis on the tropics, pp. 204–222. In: The Biology of Nectaries. Eds. Bentley, B.L. and Elias, T.S. Columbia University Press.

Bernhardt, P. 2000. Convergent evolution and adaptative radiation of beetle-pollinated angiosperms. *Plant Syst. Evol.*, **222:** 293–320.

Bernhardt, P. and Thien, L.B. 1987. Self-isolation and insect pollination in the primitive angiosperms: new evaluations of older hypothesis. *Plant Syst. Evol.*, **156:** 159–176.

Bertin, R.I. 1989. Pollination biology, pp. 23-86. In: Plant-Animal Interactions. Ed. Abrahamson, W.G. McGraw-Hill Book Company.

Bertin, R.I. and Newman, C.M. 1993. Dichogamy in angiosperms. *Bot. Rev.*, **59:** 113–153.

Bhaskar, V. and Gopinath, K. 1975. Nocturnal pollination in *Antirrhinum majus* Linnaeus by *Xylocopa rufescens* Fabricus. *J. Bombay Nat. Hist. Soc.*, **72:** 602–604.

Bhaskara Rao, C. and Subba Reddi, C. 1994a. Reproductive biology of *Cochlospermum religiosum*. *J. Trop. Ecol.*, **35:** 209–218.

Bhaskara Rao, C. and Subba Reddi, C. 1994b. Pollination ecology of *Martynia annua* L. *J. Bombay. Nat. Hist. Soc.*, **91:** 187–193.

Bhaskara Rao, C., Subba Reddi, C., Raju, A.J.S. and Atluri, J.B. 1998. Pollination ecology of *Cassia alata* L. *J. Bombay Nat. Hist. Soc.*, **95:** 454–459.

Bhatnagar, S. 1989. Plant animal interactions with special reference to strategies of growth, phenology and pollination, pp. 253–270. In: Prespectives in Ecology. Eds. Singh, J.S. and Gopal, B. Jagmandir Book Agency, New Delhi, India.

Borges, R.M., Hema, S. and Subhash, M. 1997. Alternations of sexes in a deciduous tree: temporal dioecy in *Bridelia retusa*. *Curr. Sci.*, **72(12):** 940–944.

Bowman, R.N. 1987. Cryptic self-incompatibility and the breeding system of *Clarkia unguiculata* (Onagraceae). *Amer J. Bot.*, **74:** 471–476.

Brian, M.V. and Brian, A.D. 1952. The wasp *Vespula sylvestris* Scopoli: feeding, foraging and colony development. *Trans. R. Ent. Soc.*, (London), **103:** 1–26.

Buchmann, S.L. 1983. Buzz pollination in angiosperms, pp. 73-113. In: Hand book of experimental pollination biology. Eds. Jones, C.E. and Little, R.J. Scientific and Academic Editions, Van Nostrand Reinhold Company Inc.

Bullock, S.H. 1985. Breeding systems in the flora of a tropical deciduous forest in Mexico. *Biotropica*, **17:** 287–301.

Burkill, I.H. 1906. Notes on the pollination of flowers in India. *J. Asiatic Soc. Bengal*, **2:** 511–525.

Byragi Reddy, T. and Subba Reddi, C. 1994. Pollination ecology of *Vitex negundo* (Verbenaceae). *Proc. Indian Nat. Sci. Acad.*, **B60:** 57–66.

Byragi Reddy, T. and Subba Reddi, C. 1995. Butterfly pollination of *Clerodendrum infortunatum* (Verbenaceae). *J. Bombay Nat. Hist. Soc.*, **92:** 166–173.

Byragi Reddy, T. and Subba Reddi, C. 1996. Pollination ecology of *Duranta repens* (Verbenaceae). *J. Bombay Nat. Hist. Soc.*, **93(2):** 193–201.

Charlesworth, B. and Charlesworth, D. 1978. A model for the evolution of dioecy and gynodioecy. *Amer. Nat.*, **112:** 957–997.

Chaturvedi, S.K. 1989. A new device of self pollination in *Boerhavia diffusa* L. (Nyctaginaceae) *Beitr. Biol. Pflanzen.*, **64:** 55–58.

Cook, C.D.K. 1982. Pollination mechanisms in the Hydrocharitaceae, pp. 1–15. In : Studies on aquatic plants, Eds. Symoens, J.J., Hooper, S.S. and Compere, P. *Royal Botanical Society of Belgium, Brussels.*

Corbet, S.A. 1978. Bees and nectar of *Echium vulgare*, pp. 21–30. In: The pollination of flowers by insects. Ed. Richards, A.J., Academic Press.

Courtney, S.P., Hill, C.J. and Westerman, A.1982. Pollen carried for long periods by butterflies. *Oikos*, **38:** 260–263.

Cruden, R.W. 1988. Temporal dioecism: Systematic breadth, associated traits and temporal patterns. *Bot. Gaz.*, **149:** 1–15.

Cruden, R.W. and Hermann-Parker, S. 1977. Temporal dioecism: An alternative to dioecism? *Evolution*, **31:** 863–866.

Cruden, R.W. and Hermann-Parker, S.M. 1979. Butterfly pollination of *Caesalpinia pulcherrima*, with observations on a psychophilous syndrome. *J. Ecol.*, **67:** 155–168.

Davidar, P. 1985. Ecological interactions between mistletoes and their avian pollinators in South India. *J. Bombay Nat. Hist. Soc.*, **82:** 45–60.

Davis, A.R. 1997. Pollination efficiency of insects, pp. 87–120. In: Pollen biotechnology for crop production and improvement. Eds. Shivanna, K.R. and Sawhney, V.K. Cambridge University Press.

Deodikar, G.B. and Suryanarayana, M.C. 1977. Pollination in the service of increasing farm production in India. *Ad. Pollen-spore Res.*, **2:** 1–23.

Diels, L. 1916. Kaferblumen bei den Ranales und ihre Bedeutung fur die phylogenese der Angiospermen. *Ber. Dtsch. Bot. Ges.*, **34:** 758–774.

Doreswamy, R. and Mohan Ram, H.Y. 1969. Studies on growth and flowering in axenic cultures of insectivorous plants. *Phytomorphology*, **19:** 363–371.

Eisikowitch, D., Ivri, Y. and Dafni, A. 1986. Reward partitioning in *Capparis* spp. Along Ecological gradient. *Oecologia*, **71:** 47–50.

Emms, S.K. 1993. Andromonoecy in *Zigadenus paniculatus* (Liliaceae): Spatial and temporal patterns of sex allocation. *Amer. J. Bot.*, **80:** 914–923.

Faegri, K. and Pijl, L. Van der. 1979. The principles of pollination ecology. Pergamon Press.

Flores, S. and Schemske, D.W. 1984. Dioecy and monoecy in the flora of Puerto Rico and the Virgin Islands: ecological correlates. *Biotropica*, **16:** 132–139.

Fox, J.F. 1985. Incidence of dioecy in relation to growth form, pollination and dispersal. *Oecologia*, **67:** 244–249.

Frankel, R. and Galun, E. (Eds.) 1977. Pollination mechanisms, reproduction and plant breeding. Springer-Verlag.

Freeman, D.C., McArthur, E.D., Harper, K.T. and Blauer, A.C. 1981. Influence of the environment on the floral sex ratio of monoecious plants. *Evolution*, **35:** 194–197.

Freeman, D.C., Doust, J.L., El-Keblawy, A., Miglia, K.J. and McArthur, E.D. 1997. Sexual specialisation and inbreeding avoidance in the evolution of dioecy. *Bot. Rev.*, **63:** 67–86.

Fryxell, P.A. 1957. Mode of reproduction in higher plants. *Bot. Rev.*, **23:** 135–233.

Ganders, F.R. 1979. The biology of heterostyly. *N. Z. J. Bot.*, **17:** 607–635.

Givnish, T.J. 1980. Ecological constraints on the evolution of breeding systems in seed plants: Dioecy and dispersal in gymnosperms. *Evolution*, **34:** 959–972.

Givnish, T.J. 1982. Outcrossing versus ecological constraints in the evolution of dioecy. *Amer. Nat.*, **119:** 849–865.

Gosh, M.K. 1964. Observations on the pollination in *Nechamandra alternifolia* (Roxb.). Thw. *Bull. Bot. Survey India.*, **6(2–4):** 185–187.

Gottsberger, G.1988. The reproductive biology of primitive angiosperms. *Taxon.*, **37:** 630–643.

Gottsberger, G., Gottsberger, S.I. 1988. Evolution of flower structures and pollination in Neotropical *Cassiinae* (Ceasalpiniaceae) species. *Phyton* (Austria), **28:** 293–320.

Grant, V. 1983. The systematic and geographical distribution of hawkmoth flowers in the temperate North American flora. *Bot. Gaz.*, **144:** 439–449.

Grant,V. 1994. Modes and origin of mechanical and ethological isolation in angiosperms. *Proc. Natl. Acad. Sci.*, (USA), **91:** 9–10.

Hagerup, O. 1943. Myre-Bestvning. *Bot. Tidskr.*, **46:** 116–123.

Harder, L.D. and Thomson, J.D. 1989. Evolutionary options for maximing pollen dispersal of animal pollinated plants. *Amer. Nat.*, **113:** 323–344.

Harder, L.D. and Barrett, S.C.H. 1995. Mating cost of large floral displays in hermaphrodite plants. *Nature*, **373:** 512–515.

Heard, T.A. 1999. The role of stingless bees in crop pollination. *Annu. Rev. Entomol.*, **44:** 183–206.

Herrera, C.M. 1989. Pollinator abundance, morphology, and flower visitation rate: analysis of the "quantity" component in a plant-pollinator system. *Oecologia*, **80:** 241–248.

Hessing, M.B. 1988. Geitonogamous pollination and its consequences in *Geranium caespitosum*. *Amer. J. Bot.*, **75:** 1324–1333.

Hickman, J.C. 1974. Pollination by ants: A low-energy system. *Science* , **184:** 1290–1292.

Holldobler, B. and Engel-Siegal, H. 1984. On the metapleural gland of ants. *Psyche*, **91:** 201–224.

Ibarra-Manriquez, G. and Oyama, K. 1992. Ecological correlates of reproductive traits of Mexican rain forest trees. *Amer. J. Bot.*, **79:** 383–394.

Jennerstern, O.1984. Flower visitation and pollination efficiency of some North European butterflies. *Ecologia*, **63:** 80–89.

Johnson, S.D. 1995. Observations of hawkmoth pollination in the South African orchid *Disa cooperi*. *Nord. J. Bot.*, **15:** 121–125.

Jyothi, P.V., Atluri, J.B. and Subba Reddi, C. 1990. Pollination ecology of *Moringa oleifera* (Moringaceae). *Proc. Indian Acad. Sci. (Plant Sci.)*, **100:** 33–42.

Jyothi, P.V., Atluri, J.B. and Subba Reddi, C. 1991. Pollination ecology of *Sanatlum album* (Santalaceae). *Tropical ecology* **32:** 98–104.

Kannan, P. 1978. Nectar-feeding adaptations of flowerbirds. *J. Bombay Nat. Hist. Soc.*, **75:** 1036–1050.

Kausik, S.B. and Raju, M.V.S. 1955. A contribution to the floral morphology and embryology of *Utricularia reticulata* Smith. *Proc. Indian Acad. Sci. (Plant Sci.)*., **41:** 155–166.

Kevan, P.G. 1995. Bee botany: pollination, foraging and floral calendars, pp. 113–116. In: The Asiatic Hive Bee. Ed. Kevan, P.G., Enviroquest, Ltd.

Kevan, P.G. 1997. Pollination biology and plant breeding systems, pp. 59–83. In: Pollen biotechnology for crop production and improvement, Eds. Shivanna, K.R. and Sawhney, V.K. Cambridge University Press.

Kevan, P.G. and Baker, H.G. 1999. Insects on flowers, pp. 553–584. In: Ecological Entomology, Eds. Huffaker, C.B. and Gutierrez, A.P., John Wiley and Sons, Inc.

Koul, P.A., Koul, A.K. and Hamal, I.A. 1984. Floral Biology of *Torilis leptophylla* (L.) Reichenb. f. *Proc. Indian Acad. Sci. (Plant Sci.)*, **93:** 449–454.

Kuijt, J. 1969. Biology of parasitic plants. University of California Press.

Kuruvilla, P.K. 1989. Pollination biology, seed setting and fruit setting in *Madhuca indica* (Sapotaceae). *Indian For.*, **115:** 22–28.

Ledig, F.T. 1986. Heterozygosity, heterosis and fitness in outbreeding plants. In: Soule, M. Ed. Conservation Biology. The Science of Scarcity and Diversity. Sunderland: Sinauer. 77–104.

Levin, D.A. and Berube, D.E. 1972. Phlox and Colias: The efficiency of a pollination system. Evolution **26:** 242–250.

Lewis, D. 1942. The evolution of sex in flowering plants. *Biol. Rev.*, **17:** 46–67.

Lloyd, D.G. and Yates, J.M.A. 1982. Interselection and the segregation of pollen and stigmas in hermaphrodite plants, exemplified by *Wahlenbergia albomarginata* (Campanulaceae). *Evolution,* **36:** 903–913.

Lloyd, D.G. and Webb, C.J. 1986. The evoidence of interference between the presentation of pollen and stigmas in angiosperms 1. Dichogamy. *N. Z. J. Bot.*, **24:** 135–162.

Mani, M.S. 1993. Pollination ecology and evolutionary trends in Compositae. *Indian Rev. Life Science,* **13:** 175–190.

May, P.G. 1985. Nectar uptake rates and optimal nectar concentrations of two butterfly species. *Ecologia,* **66:** 381–386.

May, P.G. 1988. Determinants of foraging profitability in two nectarivorous butterflies, *Ecol. Entomol.*, **13:** 171–184.

May, P.G. and Spears, E.E. 1988. Andromonoecy and variation in phenotypic gender of *Passiflora incarnata* (Passifloraceae). *Amer. J. Bot.*, **75:** 1830–1841.

Michener. C.D. 1974. The social behaviour of bees: A comparitive study. Belknap Press.

Mishra, R.C. 1989. Honeybee (*Apis* species) research in India: Achievements, problems and possibilities. *Ind. J. Agric. Sci.*, **59:** 85–90.

Mondal, S., Bhattacharya, K.N. and Mandal, S. 1993. Thrips-pollinator of *Peltophorum inerme* (Roxb.) Llanos. *Indian Biologist,* **25(2):** 54–57.

Mulligan, G.A. 1972. Autogamy, Allogamy and Pollination in some Canadian weeds. *Can. J. Bot.*, 50: 1767–1771.

Niklas, K.J. 1985. The aerodynamics of wind-pollination. *Bot. Rev.*, **51(3):** 328–386.

Opler, P.A. 1983. Nectar production in a tropical ecosystem, pp. 32–79. In: The Biology of Nectaries. Eds. Benteley, B. and Elias, T. Columbia University Press.

Peakall, R., Beattie, A.J. and James, S.H. 1987. Pseudocopulation of an orchid by male ants: a test of two hypothesis accounting for the rarity of ant pollination. *Oecologia* (Berlin), **73:** 522–524.

Pivnick, K.A. and Mc Neil, J.N. 1985. Effects of nectar concentration on butterfly feeding: measured feeding rates for *Thymelicus lineola* (Lepidoptera : Hesperiidae) and a general feeding model for adult Lepidoptera. *Ecologia,* **66:** 226–237.

Primack, R.B. and Lloyd, D.G. 1980. Andromonoecy in the New Zealand montane shrub manuka, *Leptospermum scoparium* (Myrtaceae). *Amer. J. Bot.*, **67:** 361–368.

Proctor, M. and Yeo, P. 1972. The pollination of flowers. Taplinger Publishing Company.

Proctor, M., Yeo, P. and Lack, A. 1996. The natural history of pollination. Timber, Portland.

Pyke, G.H. and Waser, N.M. 1981. The production of dilute nectars by humming bird and honey eater flowers. *Biotropica,* **13:** 260–270.

Raju, A.J.S., Subba Reddi, C. and Ramdas, K. 1997. Dioecism and hawkmoth pollination in *Xeromphis spinosa* Keay (Rubiaceae). *Intl. J. Ecol. Env. Sci.*, **23:** 251–260.

Raju, A.J.S., Ram Das, K., Atluri, J.B., Subba Reddi, C. and Bahadur, B. 1997. Sexual systems and pollination in distylous *Hugonia mystax* L. (Linaceae). *J. Palynol.,* **33:** 185–202.

Rama Devi, K., Atluri, J.B. and Subba Reddi, C. 1989. Pollination ecology of *Zizyphus mauritiana* (Rhamnaceae). Proc. *Indian Acad. Sci. (Plant Sci.),* **99:** 223–239.

Ramirez, N. and Brito, Y. 1990. Reproductive biology of a tropical palm swamp community in the Venezuelan Llanos. *Amer. J. Bot.,* **77(10):** 1260–1271.

Ramirez, N., Sobrevila, C., DeEnrech, N.X. and Ruiz-Zapata, T. 1984. Floral biology and breeding system of *Bauhinia benthamiana* Taub. (Leguminosae). A bat-pollinated tree in Venezuelan "Llanos". *Amer. J. Bot.,* **71:** 273–280.

Rao, S.P., Atluri, J.B. and Subba Reddi, C. 2001. Intermittent mass blooming, midnight anthesis and rockbee pollination in *Pterocarpus santalinus* (Fabaceae). *Nord. J. Bot.* **21:** 271–276.

Reddi, E.U.B. 1987. Under-pollination: A major constraint of cashewnut production. *Proc. Indian Natn. Sci. Acad.,* **B53:** 249–252.

Reddi, E.U.B. 1989. Thrips-pollination in Sapodilla (*Manilkara zapota*). *Proc. Indian Natn. Sci. Acad.,* **B55:** 407–410.

Reddi, E.U.B. and Subba Reddi, C. 1984. Pollination ecology of *Euphorbia geniculata* (Euphorbiaceae) *J. Bombay Nat. Hist. Soc.,* **81:** 571–582.

Renner, S.S. and Fiel, J.P. 1993. Pollinators of tropical dioecious angiosperms. *Amer. J. Bot.,* **80:** 1100–1107.

Renner, S.S. and Ricklefs, R.E. 1995. Dioecy and its correlates in the flowering plants. *Amer. J. Bot.,* **82:** 596–606.

Richards, A.J. 1986. Plant breeding systems. George Allen and Unuiss.

Rico-Gray, V. 1989. The importance of floral and circum-floral nectar to ants inhabiting dry tropical low lands. *Biol. J. Linn. Soc.,* **38:** 173–181.

Roubik, D.W. 1989. Ecology and natural history of tropical bees. Cambridge University Press.

Roy, R.P. 1974. Sex mechanism in higher plants. *J. Indian Bot. Soc.,* **53:** 141–155.

Ruiz, T. and Arroyo, M.T.K. 1978. Plant reproductive ecology of a secondary deciduous forest in Venezuela. *Biotropica,* **10:** 221–230.

Schmitt, J. 1980. Pollinator foraging behaviour and gene dispersal in *Senecio* (Compositae) *Evolution,* **34:** 934–943.

Sihag, R.C. 1984. Influence of environmental factors on the pollination activity of bees. *Environ. Ecol.,* **2:** 149–152.

Smith, G.R. and Snow, G. 1976. Pollination ecology of *Platanthera* (*Habenaria*) *ciliaris* and *P. blephariglottis* (Orchidaceae). *Bot. Gaz.,* **137:** 133–140.

Solbrig, O.T. 1976. On the relative advantages of cross-and self-fertilisation. *Ann. Mo. Bot. Gard.,* **63:** 260–276.

Srivastava, P.K. 1993. Pollination mechanisms in genus *Terminalia* Linn. *Indian For.,* **119:** 147–150.

Stebbins, G.L. 1950. Variation and evolution in plants. Columbia University Press.

Stebbins, G.L. 1970. Adaptive radiation of reproductive characteristics in angiosperms, I. Pollination mechanisms. *Annu. Rev. Ecol. Syst.,* **1:** 307–326.

Steiner, K.E. 1988. Dioecism and its correlates in the Cape flora of South Africa. *Amer. J. Bot.,* **75:** 1742–1754.

Subba Reddi, C. 1976. Floral mechanism, pollen productivity and pollen incidence in *Madhuca indica* Gmelin. with remarks on the mode of pollination. *New Botanist,* **3:** 11–16.

Subba Reddi, C. and Atluri, J.B. 1981. Floral biology of *Mimusops elengi* Linn. *J. Bombay Nat. Hist. Soc.,* **77:** 471–475.

Subba Reddi, C. and Reddi, E.U.B. 1984. Wind-pollination in two tropical tree species of Euphorbiaceae. *Proc. Indian Nat. Sci. Acad.,* **B50:** 60–80.

Subba Reddi, C. and Meera Bai, G. 1986. Flower-feeding by butterflies: Mutualism or parasitism. *Bull. Bot. Surv. India*, **28:** 81–88.

Subba Reddi, C., Reddi, E.U.B., Reddi, N.S. and Reddi, P.S. 1983. Reproductive ecology of *Sapindus emarginatus* Vahl (Sapindaceae). *Proc. Indian Natn. Acad.*, **49B:** 57–72.

Subba Reddi, C., Reddi, N.S. and Atluri, J.B. 1988. Circadian patterns of pollen release in some species of Poaceae. *Rev. Palaeobot. Palynol.*, **54:** 11–42.

Subba Reddi, C., Jyothi, P.V. and Atluri, J.B. 1989. Reproductive ecology of silk cotton tree *Ceiba pentandra* L. in India. *J.Palynol.*, **25:** 93–105.

Subba Reddi, C., Raju, A.J.S. and Bahadur, B. 1995. Temporal dioecism and pollination in two petro-plant species. *J. Palynol.*, **31:** 253–260.

Subba Reddi, C., Rama Das, K., Raju, A.J.S. and Atluri, J.B. 1996. Sexual system and pollination ecology of *Gmelina asiatica* L. (Verbenaceae) *J. Palynol.*, **32:** 41–50.

Subba Reddi, C., Bhaskara Rao, C. and Atluri, J.B. 1998. Floral ecology of *Gloriosa superba* (Liliaceae). *Ann. For.*, **6:** 225–231.

Subramanya, S. and Radhamani, T.R. 1993. Pollination by birds and bats. *Curr. Sci.*, **65:** 201–209.

Thien, L.B. 1980. Patterns of pollination in the primitive angiosperms. *Biotropica*, **12:** 1–13.

Thien, L.B., Azuma, H. and Kawano, S. 2000. New perspectives on the pollination biology of basal angiosperms. *Int. J. Plant Sci.*, **161:** S 225–S 235.

Thomson, J.D. and Barrett, S.C.H. 1981. Selection for outcrossing, sexual selection and the evolution of dioecy in plants. *Amer. Nat.*, **118:** 443–449.

Vogel, S. 1958. Fledermausblumen in Sudamerika. *Osterr. Bot. Zeitschr.*, **104:** 491–530.

Watanabe, M.1992. Egg maturation in laboratory reared females of the swallow tail butterfly, *Papilio xuthus* L. (Lepidoptera : Papilionidae), feeding on different concentration solutions of sugar. *Zool. Sci.*, **9:** 133–141.

Weller, S.G. and Ornduff, R. 1977. Cryptic self-incompatibility in *Amsinckia grandiflora*. *Evolution*, **31:** 47–51.

Wesley, H.D. 1987. Bird activity and seed productivity in the coral tree, *Erythrina Indica*. *Indian For.*, **113:** 640–647.

Westerkamp, C. 1990. Bird-flowers: hovering versus perching exploitation. *Bot. Acta.*, **103:** 366–371.

Wiklund, C., Eriksson, T. and Lundberg, H. 1979. The wood white butterfly *Leptidea sinapis* and its nectar plants: A case of mutualism or parasitism. *Oikos*, **43:** 358–362.

Willson, M.F. 1979. Sexual selection in plants. *Amer. Nat.*, **113:** 777–790.

Willson, M.F. 1983. Plant reproductive ecology. John Wiley.

Wyatt, R. 1983. Pollinator-plant interactions and the evolution of breeding systems, pp. 51–95. In: Pollination biology, Ed. Real, L. Academic Press.

Yampolsky, C. and Yampolsky, H. 1922. Distribution of sex forms in the phanerogamic flora. *Bibl. Genet*, **3:** 1–62.

CHAPTER 6

GENUS *HABENARIA* WILLD. (ORCHIDACEAE) IN MAHARASHTRA

M.M. Sardesai and S.R. Yadav***

ABSTRACT

The present paper reports on an occurrence of 22 species and 3 varieties of *Habenaria* in the State of Maharashtra and provides botanical key to the identification, distribution and ecological notes.

Introduction

Family Orchidaceae, one of the largest families of flowering plants, is cosmopolitan in distribution. It has attracted an attention of number of botanists, naturalists and enthusiasts due to its peculiar incredible shapes and colours of the flowers. It is represented by about 1141 species, 166 genera widely distributed in India (Sathish Kumar and Manilal, 1994). *Habenaria* Willd., mainly the terrestrial, tuberous genus (rarely lithophytic — *Habenaria crinifera* Lindl.) is represented by about 800 species distributed throughout the world especially tropics and temperate regions of the world. In India it is represented by 72 species, of 38 species are endemic. The present paper reports an occurrence of 22 species and 3 varieties of *Habenaria* found in Maharashtra and provide botanical key to the identification, distribution and ecological notes for the species.

* Dept. of Botany, Shivaji University, Kolhapur-416 004

** Dept. of Botany, Delhi University, Delhi-110 007

Habenaria commelinifolia

Habenaria crassifolia

Habenaria crinifera

Habenaria digitata

Plate I

Habenaria foliosa var. *foetida*

Habenaria foliosa var. *foliosa*

Habenaria foliosa var. *gibsonii*

Habenaria furcifera

Plate II

Habenaria grandifloriformis

Harbenaria heyneana

Habenaria longicorniculata

Habenaria marginata

Plate III

Habenaria multicaudata

Habenaria ovalifolia

Habenaria panchganiensis

Habenaria perrottetiana

Plate IV

Habenaria plantaginea

Habenaria rariflora

Habenaria roxburghii

Habenaria stenopetala

Plate V

Topography

Maharashtra State is situated between 22° 1′ to 16° 4′ north latitudes and 72° 6′ and 80° 9′ east longitudes encompassing an area of about 3,07,690 sq. kms. The maximum height reaches to 1660 meters. East-West spread of the district is about 800 kms and across the North-South is about 700 kms. It is bordered by Gujrat and Madhya Pradesh on north and Andhra Pradesh on east. It meets the Karnataka on east and south and Goa on south-west. The Arabian Sea on west forms natural boundary of the State.

The area shows marked differences in topographic features. The Western Ghats — one of the 'Hotspots', is main range of gently uneven and mature looking crestline of 'Sahyadris'. The Sahyadris lies between 15° 60′ to 20° 75′ north latitudes and 72° 61′ and 74° 40′ east longitudes traverse 11 districts. The main range of Sahyadris divide State into two regions, i.e., Konkan and Plateau. Thus the State of Maharashtra can be divided into (i) Konkan — a narrow strip running parallel to Sahyadris, about 27–48 kms. broad and 800 kms. long. (ii) Sahyadris — the main range of gently uneven and mature looking crestline with a height between 800 and 1650 meters and a width of 15 to 80 km, spread in a north-south direction in the length of 750 kms., (ii) The plateau — the rest of the Maharashtra State is marked by several hill ranges which emerge from the main range and develop an eastward trend gently undulating with ridges and valleys. They have a height between 600 and 800 meters and a well-developed drainage pattern. The tributaries of the River Bhima, Gadnandi, Godavari, Krishna, Narmada, Purna, Shastri, Shuk, Tapi, Terekhol, Vainganga, Vaitarana, Vasishisthi, and Wardha drain the area through the State to the Arabian Sea and Bay of Bengal.

Review of Literature

In number of floristic works, the species of *Habenaria* are poorly represented because they escape from the attention of workers due to their short life spans.

With the help of native physicians and people, van Rheede studied plants of Malabar Coast. This great collaborative work was published under the title, 'Hortus Indicus Malabaricus' in 12 volumes, which provide 793 illustrations covering about 691 taxa (1678–1693). Linnaeus published 'Species Plantarum', which includes some orchids, all of which are treated under '*Epidendrum*', but interestingly no single species of *Habenaria*. John Lindley, father of ochidology contributed greatly to the Indian orchids (1753 and 1759). Roxburgh studied the flora of Madras Coast and published 'Flora Indica' (1820 and 1824).

Graham collected 1500 species from Khandala and other regions of the ghats. It was through the collaboration of various plant collectors, Graham and Nimmo could publish 'A Catalogue of Plants Growing in Bombay and it's vicinity' (1839). Nicol Alexander Dalzell prepared short descriptions of the plants in Graham's Catalogue and thus came out with 'Bombay Flora' (Dalzell and Gibson, 1861), which reports 13 species of *Habenaria*. Of these 13 species *H. caranjensis* and *H. suaveolens* are described by Dalzell (1850).

Theodore Cooke arrived in India in 1860 as a railway Engineer and studied plants of the Bombay Presidency with the help of D. Ritchiei, G.M. Woodrow, J.E. Stock, J.E. Shaw, N.A. Dalzell, A. Gibson, P.S. Kanitkar, L.D. Garade and others. He published 'The Flora of the Presidency of Bombay' (Cooke, 1901–1908) covering plants of Sind, Gujrat, Maharahstra and North Karnataka reporting just 10 species of *Habenaria* for the Bombay Presidency including *H. caranjensis* Dalz. Gammie (1905–1912) also contributed to the orchids of Bombay.

Rev. Ethelbert Blatter and Charles McCann (1931–1932) did great deal of field work. They described many species for the Bombay. Rev. Hermengild Santapau did extensive work on plants of presidency of Bombay and published number of important books. He reported 10 species and 1 variety for the Khandala (1953) and 11 species for the Purandhar (1958) including one nomenclature change, i.e., *H. hollandiana*. *Habenaria panchganiensis* Sant. & Kapad. (1957) is described by him. Santapau and Kapadia (1966) in orchids of Bombay enumerated 21 species, 3 varieties and 1 forma excluding *H. caranjensis* Dalz.

In the last four decades, number of floras have been published on different regions and districts of Maharashtra. Vartak (1966) in 'Enumeration of plants from Gomantak, India' by enumerating only 2 species. The Following regional and district floras viz 'Flora of Melghat' (Patel, 1968) with 4 species, 'Flora of Osmanabad' (Naik, 1979) with 2 species, 'Flora of Nagpur' (Ugemuge, 1986) with single species, Akola (Kamble and Pradhan, 1988) with 6 species, Sindhudurg (Kulkarni, 1988) with 6 species, 'Flora of Sawantwadi' (Almeida, 1990) with 12 species, Nasik (Lakshminarasimhan and Sharma, 1991) with 10 species, Raigad (Kothari and Moorthy, 1993) with 9 species, Mahabaleshwar-Satara (Deshpande *et al.*, 1995) with 12 species and 1 variety, Yevatmal (Karthikeyan and Anandkumar, 1993) with 2 species of *Habenaria*.

Karthikeyan *et al.* (1981) in his 'Catalogue of species added to Cooke's Flora of Bombay Presidency' added 8 species and 5 varieties for the State of Maharashtra. Karthikeyan (1989) enumerated 69 species and 8 varieties and 1 forma for India, out of which 22 species and 3 varieties occur in the State. Sharma *et al.* (1996) enumerated 21 species and 3 varieties for Maharashtra State. Naik *et al.* (1998) worked out 'Flora of Marathwada', which reports 4 species of *Habenaria* for the Marathwada region, which includes 7 districts such as Aurangabad, Beed, Jalna, Latur, Nanded, Osmanabad and Parbhani. 19 species and 3 varieties are reported for the Kolhapur District (Sardesai and Yadav, 2000 and Yadav and Sardesai, 2002).

Materials and Methods

The present work is based on the collections made by authors, available herbarium specimens and published literature. Plant materials were collected from various localities during last fifteen years. Special attention was paid to under or unexplored remote areas of the State. As far as possible, all the important localities have been covered during all the seasons. While collecting the specimens, the data such as habit, habitat, fragrance and colour of flower, frequency of distribution, local names and uses were recorded in field itself and field number was given to each specimen. All the specimens were processed for drying by using regular drying method with blotting and newspapers.

Most of the specimens and field identifications were confirmed satisfactorily with the help of regional and national floras and available literature. Identifications of the specimens were confirmed by matching them with authentically identified specimens in the Herbarium of Botanical Survey of India, Western Circle, Pune and Blatter Herbarium, St. Xavier's College, Mumbai.

After proper processing and poisoning, the specimens were mounted on herbarium sheets. Each species is well preserved in the herbarium. Critical observations were made in field as well as in laboratory. Based on observations, a key is provided for correct identification of the species. Description, distribution, ecological note and herbarium specimens collected are recorded under each species. The species are arranged alphabetically in the text. The voucher specimens of genus *Habenaria* are deposited in Herbarium (SUK) of Shivaji University, Kolhapur. The species are arranged alphabetically. For each species correct name has been cited. The nomenclature of all taxa has been made up to date as far as possible.

Taxonomic Account

Botanical keys based on various characters for identification of species of *Habenaria* are given below.

Key to Species

1. Petals two partite:
  2. Sepals without filiform tips; lobes of lip subequal or lateral lobes much longer than midlobe:
    3. Leaves scattered along stem or at the base of stem:
      4. Leaves few, radical; flowers white, faintly scented:
        5. Leaves oblong-lanceolate, thin, not flat on ground; anthers rounded at tip .... ***H. rariflora***
        5. Leaves ovate to almost orbicular, fleshy, flat on ground; anthers distinctly tricuspidate at tip .... ***H. grandifloriformis***
      4. Leaves scattered along stem; flowers greenish, fowl scented:
        6. Flowers pure green, strongly fowl scented in the night time; lower segments of petals filiform, less than half as long as upper ones; spur scarcely clavate .... ***H. digitata***
        6. Flowers pure white or greenish white; odorless or fowl scented in day time; segments of petals subequal, not filiform; spur distinctly clavate at the base:
          7. Sepals atleast green; petals white or greenish white
            8. Plants robust; sepals green, petals white; flowers scentless during day time.......... ***H. foliosa*** var. ***foetida***
            8. Plants delicate; sepals and petals greenish-white .... ***H. foliosa*** var. ***foliosa***
          7. Sepals and petals pure white .... ***H. foliosa*** var. ***gibsoni***
    3. Leaves clustered at the middle of stem ....***H. multicaudata***
  2. Sepals with filiform tips; lateral lobes of lip less than half as long as midlobe .... ***H. stenopetala***
1. Petals entire:
  9. Leaves ovate-orbicular, flat on ground
    10. Lip longer than lateral sepals, three partite; lobes subequal or lateral ones longer; spur about equaling the ovary; tubes of anther cells short:
      11. Flowers greenish white; petals narrow, linear-oblong; segments of lip filiform, lateral ones much longer than midlobe ....***H. diphylla***
      11. Flowers white; petals ovate or ovate-oblong; segments of lip subequal, linear-oblong .... ***H. crassifolia***
    10. Lip shorter than lateral sepals, three lobed; lateral lobes oblong-rounded, half as long as the whole lip; spur twice as long as the ovary; tubes of anther cells long .... ***H. roxburghii***
  9. Leaves not as above:
    12. Petals with filiform tails:
      13. Ovary long beaked; petals broad-oblong; lip three partite, segments filiform, lateral ones longer; tubes of anther cells long ....***H. commelinifolia***
      13. Ovary scarcely beaked; petals narrow linear; lip three partite, lateral lobes of lip with filiform tails, midlobe two lobulate with minute mucro between filiform tailed lobules; tubes of anther cells short .... ***H. crinifera***
    12. Petals and lobes of lip without filiform tails:

14. Lateral lobes of lip not broader than midlobe:
    15. Leaves radical, clustered at the base or middle of scape:
        16. Leaves clustered about middle of scape; flowers green
            17. Midlobe of lip inflexed; spur straight .... ***H. ovalifolia***
            17. Midlobe of lip straight; spur incurved:
                18. Spur longer than ovary .... ***H. furcifera***
                18. Spur shorter than ovary .... ***H. hollandiana***
        16. Leaves radical, clustered at the base of scape; flowers yellow or green or white, not secund:
            19. Leaves narrow lanceolate or linear-oblong; floral bracts up to half as long as ovary .... ***H. viridiflora***
            19. Leaves broad, oblong; floral bracts equal to or longer than ovary.. ***H. marginata*** var. ***marginata***
    15. Leaves scattered along stem:
        20. Flowers white: .... ***H. heyneana***
        20. Flowers yellow:
            21. Flowers secund in one or two rows .... ***H. perrottetiana***
            21. Flowers not secund .... ***H. marginata*** var. ***flavescens***
14. Lateral lobes of lip broader than midlobe:
    22. Spur 2 or 3 times longer than ovary
        23. Leaves radical, ovary sessile or subsessile; stigmatic processes somewhat oblong, white .... ***H. plantaginea***
        23. Leaves clustered at the base, ovary long pedicelled; stigmatic processes clavate, green .... ***H. longicorniculata***
    22. Spur shorter than or equaling ovary:
        24. Lip as long as or shorter than the lateral sepals; spur about as long as ovary .... ***H. suaveolens***
        24. Lip more than 1.5 times longer than that of lateral sepals; spur much shorter than ovary .... ***H. panchganiensis***

1. **Habenaria commelinifolia** (Roxb.) Wall. ex Lindl. Gen. Sp. Orchid. 325. 1835; Hook.f. Fl. Brit. India 6: 143. 1890; Cooke, Fl. Pres. Bombay 3: 224. 1958 (Repr. ed.); Sant. and Kapad. Orch. Bombay 25, t. 4, f. 11–12. 1966; Lakshmi. in Sharma *et al.* (ed.) Fl. Maharashtra St. Monocotyledones 34. 1996. *Orchis commelinifolia* Roxb. Fl. Ind. 3: 415. 1832.

Tuberous, perennial, robust herbs, ca 90 cm tall. Tubers 1 to 2, ellipsoidal, covered with hairs. Stem stout, erect, terete with 4 to 5 sheaths towards base; sheaths infundibuliform, oblique at mouth. Leaves 6–7 gradually changing into floral bracts, distichous, conduplicate with sheathing base, 5–25 to 1.5–3 cm, oblong-oblanceolate, acute at apex, margins hyaline, minutely papillate, 3–5-nerved. Flowers white, sessile or subsessile, bracteate in 20–40 cm long, many flowered dense spikes. Bracts 4–5 cm long, acuminate, unequal, finely ciliolate along margins. Sepals unequal, white; dorsal sepal 6.5–7 × 5–5.5 mm, erect, obovate-oblong, mucronulate at tip, 3-nerved. Lateral sepals 12–13 × 6.5–7 mm, narrower than dorsal sepal and lateral petals, obovate-oblong, oblique at base, 3-nerved. Petals white, 5.5–6 × 3.5–4 mm, falcate, erect, broadly oblique towards base, acute at tip, 1-nerved. Lip white, 3 cm long; thick, spurred, tripartite almost to the base, clawed; lobes unequal, spreading; lateral lobes 2–3 cm, longer than midlobe, acute, curved; midlobe 1.5–2 cm long, narrowly linear. Spur 5–7 cm long, clavate, much longer than ovary. Ovary 3–3.5 cm long, curved, sessile or subsessile, beaked; beak ca 2 cm long. Capsule 4–4.5 cm long, ribbed.

Illus.: Sant. and Kapad. *op. cit.*

Fls. and Frts.: Aug.–Dec. *Exsiccata*: MMS 2039

Occurrence: Frequent.

Distribution: Amaravati, Bhandara, Chandrapur, Gadchiroli, Gondiya, Kolhapur, Mumbai, Nashik, Pune, Raigad, Ratnagiri, Sindhudurg and Thane districts of Maharashtra.

Ecological Note: Grow on hills slopes in dry deciduous forests.

Taxonomic Note: Robust herbs. Petals with long filiform appendages is the characteristic of this species.

2. **Habenaria crassifolia** A. Rich. in Ann. Sci. Nat. 2,15: 72, t. 36. 1841; Hook.f. Fl. Brit. India 6: 151. 1890; Cooke, Fl. Pres. Bombay 3: 227. 1958 (Repr. ed.); Sant. and Kapad. Orch. Bombay 22, t. 32. 1966; Lakshmi. in Sharma *et al.* (ed.) Fl. Maharashtra St. Monocotyledones 35. 1996.

Tuberous, perennial herbs, 20–50 cm tall. Tubers 1 to 2, unequal, ovate-oblong. Stem slender, erect, with many sheaths towards base. Leaves 2 or rarely 3, somewhat fleshy, lying flat on the ground, distichous with sheathing base, very variable, 2–4 × 1.8–6 cm, oblong-obovate, abruptly acuminate at apex, margins hyaline, minutely papillate. Flowers white, sessile or subsessile, bracteate in 20–50 cm long, many flowered lax spikes. Bracts 2–5 cm long, acuminate, unequal, 1-nerved. Sepals unequal, white; dorsal sepal 4 × 3.5 mm, erect, ovate-orbicular, mucronulate at tip, 3-nerved. Lateral sepals 6 × 4 mm, broader than dorsal sepal and lateral petals, ovate-oblong, oblique at base, 3-nerved. Petals white, 2.8–3 × 2–2.5 mm, erect, ovate or oblong-obovate, oblique towards base, acute at tip, 1-nerved. Lip white, 5 mm long; thick, spurred, tripartite little above the middle; lobes subequal, linear, spreading, reflexed; lateral lobes acute; midlobe obtuse. Spur 1.5–2 cm long, clavate, as long as or shorter than ovary. Ovary 2–2.5 cm long, curved, sessile or subsessile. Capsule 2–3 cm long, ribbed.

Illus.: Sant. and Kapad. *op. cit.*

Fls. and Frts.: Aug.–Dec. *Exsiccata*: MMS 165, MPB 5205

Occurrence: Rare.

Distribution: Ahmednagar, Kolhapur, Nashik, Pune, Raigad, Ratnagiri, Sangali, Satara, Sindhudurg and Thane districts of Maharashtra.

Ecological Note: Grow amidst grasses at high altitudes over 900 meters. Flowers little fragrant.

Taxonomic Note: Leaves flat on the ground. Flowers white; lip 3-lobed, lobes equal.

3. **Habenaria crinifera** Lindl., Gen. Sp. Orchid. 323. 1835; Hook.f. Fl. Brit. India 6:. 142. 1890; Cooke, Fl. Pres. Bombay 3: 223. 1958 (Repr. ed.); Sant. and Kapad. Orch. Bombay 26, t.4, f. 13–14. 1966; Lakshmi. in Sharma *et al.* (ed.) Fl. Maharashtra St. Monocotyledones 35. 1996.

Tuberous, perennial, lithophytic, pendulous herbs. Tuber single, ovoid. Stem slender, pendulous, terete, ca 30 cm tall with 2 to 3 sheaths towards base. Leaves 2–7, distichously clustered at base with sheathing base, 5–15 to 0.5–2 cm, oblong-oblanceolate, acute at apex, margins hyaline. Flowers 2–8, white, shortly pedicellate, bracteate in 5–25 cm long racemes. Bracts 1–2 cm long, acuminate, always shorter than ovary. Sepals subequal, greenish-white; dorsal sepal 7–8 × 7–8 mm, erect, obovate-oblong, obtuse at tip, 3-nerved. Lateral sepals 10–11 × 6.5–7 mm, narrower than dorsal sepal, oblong, oblique at base, 3-nerved. Petals white, 7–8 × 1.5–2 mm, linear, erect, acute at tip, 1-nerved. Lip white, 1.5–2 cm long, spurred, trilobed, clawed; lobes unequal, spreading; lateral lobes 1.5–2 cm, longer than midlobe, acuminate, prolonged into a filiform tail; midlobe 1–1.2 cm long, 2-lobulate. Spur 3.5–4 cm long, clavate, greenish-white, longer than ovary. Ovary 2–2.5 cm long, curved, subsessile, beaked; beak ca 5–7 mm long. Capsule 2.5–3 cm long, ribbed.

Illus.: Sant. and Kapad. *op. cit.*; Abraham and Vatsala Introduction to Orchids f. 240. 1981.

Fls. and Frts.: Aug.–Dec. *Exsiccata*: MMS 163, MPB 5968

Occurrence: Rare.

Distribution: Kolhapur, Ratnagiri, Sangali, Satara and Sindhudurg districts of Maharashtra.

Ecological Note: Lithophytic on crevices of trees at higher altitudes. Very rarely terrestrial. Flowers fragrant. It also propagates vegetatively by formation of tubers on lateral roots.

Taxonomic Note: It is the only lithophytic species. Delicate herbs. Petals with filiform appendages.

4. **Habenaria digitata** Lindl. Gen. Sp. Orchid. 307. 1835; Hook. f. Fl. Brit. India 6: 134. 1890; Cooke, Fl. Pres. Bombay 3: 220. 1958 (Repr. ed.); Sant. and Kapad. Orch. Bombay 10, t.3, f. 7–8. 1966; Lakshmi. in Sharma *et al.* (ed.) Fl. Maharashtra St. Monocotyledones 35. 1996.

Tuberous, perennial, robust herbs, ca 40 cm tall. Tubers 1 to 2, elliptic-ovoid. Stem stout, erect, terete with 3 to 5 sheaths towards base. Leaves 6–7 gradually changing into floral bracts, pentastichous with sheathing base, 2–15 × 1.5–5 cm, obovate-oblong, acute at apex, margins hyaline, strongly 3-nerved. Flowers green or tinged with yellow or white, sessile or subsessile, bracteate in 5–15 cm long, many flowered dense spikes. Bracts 2.5–3 cm long, acuminate. Sepals subequal, green; dorsal sepal 7–9 × 6–7 mm, erect, obovate, 3-nerved. Lateral sepals 10–13 × 5–6 mm, narrower than dorsal sepal, obovate-oblong, oblique at base, 3-nerved. Petals green, 2-partite, 7–8 × 1–2 mm, subfalcate, erect, broadly oblique towards base, acute at tip, 3-nerved. Lip green, 1.5–2 cm long; thick, spurred, tripartite almost to the base; lobes subequal, spreading; lateral lobes acute, spreading at right angle to the column; midlobe, narrowly linear, straight or inwardly reflexed. Spur green, 1.5–2 cm long, clavate, as long as or little shorter than ovary. Ovary 1.5–2 cm long, curved, sessile or subsessile, beaked. Capsule 2–2.5 cm long, ribbed.

Illus.: Sant. and Kapad. *op. cit.*

Fls. and Frts.: Aug.–Dec. *Exsiccata*: MMS 148, MPB 6003

Occurrence: Frequent.

Distribution: Ahmednagar, Kolhapur, Mumbai, Nashik, Pune, Raigad, Ratnagiri, Sangali, Satara, Sindhudurg and Thane districts of Maharashtra.

Ecological Note: Grow amidst grasses, very strongly fowl scented in the evening. Entire flower greenish.

Taxonomic Note: Robust herbs. Flowers pure green, strongly fowl scented in the night time. Petals 2-partite; lower segments of petals filiform, less than half as long as upper ones.

5. **Habenaria diphylla** Dalz. In Hook. Kew J. Bot. 2: 262. 1850; Hook. f. Fl. Brit. India 6: 151. 1890; Cooke, Fl. Pres. Bombay 3: 227. 1958 (Repr. ed.); Sant. and Kapad. Orch. Bombay 21, t. 6, f. 22. 1966; Lakshmi. in Sharma *et al.* (ed.) Fl. Maharashtra St. Monocotyledones 36. 1996.

Tuberous, perennial herbs, 15–30 cm tall. Tubers 1 to 2, unequal, ovate-oblong. Stem slender, erect, with 2–3 sheaths towards base. Leaves 2 or rarely 3, somewhat fleshy, lying flat on the ground, distichous with sheathing base, very variable, 1.5–4 × 0.8–4 cm, oblong-obovate, acute or obtuse at apex, margins hyaline. Flowers white or greenish-white, shortly pedicellate, bracteate in 15–30 cm long, many flowered dense racemes. Bracts 0.8–1 × 0.5–0.7 cm long, acute, 1-nerved. Sepals subequal, greenish-white; dorsal sepal 5–7 × 4–5 mm, erect, ovate, mucronulate at tip, 3-nerved. Lateral sepals 6–9 × 4–5 mm, broader than dorsal sepal, ovate-oblong, oblique at base, 3-nerved. Petals white, 5–7 × 2–3 mm, erect, subfalcate, linear or oblong-obovate, oblique towards base, acute at tip, 1-nerved. Lip white, 1.5–1.8 mm long; thick, spurred, trilobed; lobes unequal, spreading, reflexed; lateral lobes acute; midlobe obtuse. Spur 1.5–2 cm long, clavate, as long as or longer than ovary. Ovary 1.5–2 cm long, curved, sessile or subsessile. Capsule 1.8–2 cm long, ribbed.

Illus.: Sant. and Kapad. *op. cit.*

Fls. and Frts.: Aug.–Dec. *Exsiccata*: MMS 68

Occurrence: Rare.

Distribution: Kolhapur, Ratnagiri and Sindhudurg districts of Maharashtra.

Ecological Note: Grow amidst grasses at lower altitudes in moist and dry deciduous forests.

Taxonomic Note: Leaves flat on ground. Flowers white or greenish-white. Lip longer than lateral sepals, three partite; lobes subequal or lateral ones longer segments of lip.

6. **Habenaria foliosa** A. Rich. var. **foliosa** A. Rich. in Ann. Sci. Nat. 2, 15: 71, t. 3A. 1841; Lakshmi. in Sharma *et al.* (ed.) Fl. Maharashtra St. Monocotyledones 37. 1996. *H. digitata* var. *foliosa* (A. Rich.) Hook. f. Fl. Brit. India 6: 135. 1890; Cooke, Fl. Pres. Bombay 3: 220. 1958 (Repr. ed.). *H. gibsonii* var. *foliosa* (A. Rich.) Sant. and Kapad. in J. Bombay nat. Hist. Soc. 56: 194, t. 2, f. 6. 1959 and Orch. Bombay 14. t. 2, f. 6. 1966.

Tuberous, perennial, delicate herbs, ca 15 cm tall. Tuber single, ellipsoid-ovoid. Stem erect, terete with 2–3 sheaths towards base. Leaves 3–5, gradually changing into floral bracts, pentastichous with sheathing base, 3.5–5 × 1.5–2 cm, obovate-oblong, acute at apex, margins hyaline, wavy, strongly 3-nerved. Flowers 3–7, white or tinged with green, sessile or subsessile, bracteate in 2.5–5 cm long, lax, secund or subsecund spikes. Bracts 2.5–3 cm long, acuminate, pale green. Sepals unequal, green; dorsal sepal 5–7 × 3–5 mm, erect, obovate, 3-nerved. Lateral sepals 0.7–1 × 0.2–0.3 cm, narrower than dorsal sepal, oblique at base, 3-nerved. Petals white or tinged with green, 2-partite almost to the base; upper segments sickle shaped, 0.5–0.8 mm, lower segment 0.4–0.7 cm, narrower, 3-nerved. Lip white, 3-partite nearly to the base, 1.5–2 cm long; spurred; lobes unequal, spreading; lateral lobes acute; midlobe longer than lateral lobes, narrowly linear. Spur white, 1.5–2 cm long, clavate, longer than ovary. Ovary 1–1.2 cm long, curved, sessile or subsessile, beaked. Capsule 1–1.5 cm long, ribbed.

Illus.: Sant. and Kapad. *op. cit.*

Fls. and Frts.: Aug.–Dec. *Exsiccata*: MMS 1485, MPB 5956

Occurrence: Rare.

Distribution: Kolhapur, Pune, Sangali and Satara districts of Maharashtra.

Ecological Note: Very delicate, grows as undergrowth in *Carvia callosa* at higher altitudes on hill slopes. This delicate variety is late flowering and little scented.

Taxonomic Note: Plants delicate; sepals and petals greenish-white; odorless; segments of petals subequal, not filiform; spur distinctly clavate at the base.

7. **Habenaria foliosa** A. Rich. var. **foetida** (Blatt. et McCann) Bennet in J. Econ. Tax. Bot. 5: 452. 1984. *H. gibsonii* var *foetida* Blatt. et McCann in J Bombay nat. Hist. Soc. 36: 16. 1932; Santapau and Kapad. in J. Bombay nat. Hist. Soc. 36: 16. 1932 and Orch. Bombay 13, t. 2, f. 5. 1966; Lakshmi. in Sharma *et al.* (ed.) Fl. Maharashtra St. Monocotyledones 37. 1996.

Tuberous, perennial, robust herbs, 15–30 cm tall. Tuber 1 or 2, ellipsoid. Stem stout, erect, terete with 3–5 sheaths towards base. Leaves 3–7, gradually changing into floral bracts, pentastichous with sheathing base, 5–15 × 2–5 cm, obovate, acute at apex, margins hyaline, wavy, strongly 3-nerved. Flowers white or tinged with green, sessile or subsessile, bracteate in 5–10 cm long, few to many flowered lax, secund spikes. Bracts 3–5 cm long, acuminate, pale green. Sepals unequal, green; dorsal sepal 7–10 × 5–7 mm, erect, obovate, 3-nerved. Lateral sepals 0.9–1.4 × 0.5–0.7 cm, narrower than dorsal sepal, oblique at base, 3-nerved. Petals white or tinged with green, 2-partite almost to the base; upper segments sickle shaped, 0.7–0.8 mm, lower segment 0.5–0.7 cm, narrower, 3-nerved. Lip white, 3-partite nearly to the base, 1.5–2 cm long; spurred; lobes unequal, spreading; lateral lobes acute;

midlobe longer than lateral lobes, narrowly linear. Spur white, 2–2.5 cm long, clavate, longer than ovary. Ovary 2–2.5 cm long, curved, sessile or subsessile, beaked. Capsule 2–2.5 cm long, ribbed.

Illus.: Sant. and Kapad. *op. cit.*

Fls. and Frts.: Aug.–Dec. *Exsiccata*: MMS 1555

Occurrence: Fairly common.

Distribution: Amaravati, Aurangabad, Kolhapur, Mumbai, Nanded, Nashik, Pune, Raigad, Ratnagiri, Sangali and Satara districts of Maharashtra.

Ecological Note: Grows on humus rich soil, as undergrowth, flowers strongly fowl scented in the daytime.

Taxonomic Note: Plants robust. Leaves strongly 3-nerved. Sepals green and petals white; segments of petals subequal, not filiform; spur distinctly clavate at the base. Fowl scented during the day time.

8. **Habenaria foliosa** A. Rich. var. **gibsonii** (Hook. f.) Bennet in J. Econ. Tax. Bot. 5: 452. 1984; Lakshmi. in Sharma *et al.* (ed.) Fl. Maharashtra St. Monocotyledones 37. 1996. *H. gibsonii* Hook. f. Fl. Brit. India 6: 135. 1890; Sant. and Kapad. Orch. Bombay 12, t. 2, f. 3–4. 1966

Tuberous, perennial herbs, 15–30 cm tall. Tuber single, ovoid. Stem erect, terete with 2–3 sheaths towards base. Leaves 3–5, gradually changing into floral bracts, pentastichous with sheathing base, 3.5–4 × 1.5–2 cm, oblong-oblanceolate, acute at apex, margins hyaline, wavy, strongly 3-nerved. Flowers 3–5, white, sessile or subsessile, bracteate in 5–7 cm long, lax secund spikes. Bracts 3–5 cm long, acuminate, pale green. Sepals unequal, green; dorsal sepal 1–1.5 × 0.8–1 cm, erect, ovate-elliptic, 3-nerved. Lateral sepals 1.7–2 × 0.8–1 cm, narrower than dorsal sepal, oblique at base, 2-nerved. Petals white, 2-partite almost to the base; upper segments sickle shaped, 0.8–1.2 cm, lower segment 0.7–1 cm, narrower, 3-nerved. Lip white, 3-partite nearly to the base, 2–2.5 cm long; spurred; lobes unequal, spreading; lateral lobes acute; midlobe longer than lateral lobes, narrowly linear. Spur white, 2–2.5 cm long, clavate, longer than ovary. Ovary 1.5–2 cm long, curved, sessile or subsessile, beaked. Capsule 2–2.5 cm long, ribbed.

Illus.: Sant. and Kapad. *op. cit.*

Fls. and Frts.: Aug.–Dec. *Exsiccata*: 173, MPB 5199

Occurrence: Frequent.

Distribution: Ahmednagar, Akola, Kolhapur, Nashik, Pune, Raigad, Ratnagiri, Sangali, Satara, Sindhudurg and Thane districts of Maharashtra.

Ecological Note: Grow on steep Western slopes of main range of Sahyadris in loose soil as undergrowth in *Carvia callosa*. Flowers sweet scented.

Taxonomic Note: Robust herbs. Leaves strongly 3-nerved. Flowers pure white, fowl scented in daytime; segments of petals unequal, not filiform; spur distinctly clavate at the base. Sepals and petals pure white.

9. **Habenaria furcifera** Lindl. Gen. Sp. Orchid. 319. 1835; Hook. f. Fl. Brit. India 6: 149. 1890; Santapau and Kapad. Orch. Bombay 39, t. 5, f. 20. 1966; Lakshmi. in Sharma *et al.* (ed.) Fl. Maharashtra St. Monocotyledones 37. 1996.

Tuberous, perennial herbs, 15–60 cm tall. Tubers 1 or 2, ovoid-ellipsoid. Stem stout, erect, terete with 3–5 sheaths towards base. Leaves 3–7, clustered at base with sheathing base, 15–20 × 3–5 cm, obovate, acute at apex, margins wavy, strongly 5–7-nerved. Flowers green, sessile or subsessile, bracteate in 15–50 cm long, many flowered lax stout spikes. Bracts 1–1.3 cm long, acuminate, pale green. Sepals subequal, green; dorsal sepal 4–5 × 0.8–1 cm, obovate, subfalcate, concave at base, 3-nerved. Lateral sepals 4–5 × 0.7–1 cm, oblique at base, 3-nerved. Petals green, 1-nerved. Lip green, 3-partite nearly to the base; spurred; lobes unequal, spreading; lateral lobes filiform, acute, diverging; midlobe shorter than lateral lobes, blunt, stout. Spur green, clavate, as long as or longer than ovary. Ovary 1.2–1.5 cm long, sessile or subsessile, beaked. Capsule ca 1.5 cm long, ribbed.

Illus.: Sant. and Kapad. *op. cit.*

Fls. and Frts.: Aug.–Dec. *Exsiccata*: MMS 1553

Occurrence: Rare.

Distribution: Ahmednagar, Kolhapur, Nashik, Ratnagiri, Sangali, Satara and Sindhudurg districts of Maharashtra.

Ecological Note: Grows in grasses on hill slopes.

Taxonomic Note: Stout herbs. Leaves radical, clustered at the base. Flowers green. Spur incurved.

10. **Habenaria grandifloriformis** Blatt. & McC. in J. Bombay nat. Hist. Soc. 36: 17. 1932 emend. Sant. and Kapad. *Ibid.* 56: 195. 1959 and Orch. Bombay 17, t. 1, f. 2–2′ 1966; Lakshmi. in Sharma *et al.* (ed.) Fl. Maharashtra St. 1: 38. 1996. *H. grandiflora* Lindl. ex Dalz. & Gibs. Bombay Fl. 267. 1861 non Torr. ex Beck. 1823; Hook. f. Fl. Brit. India 6: 136. 1890; Cooke, Fl. Pres. Bombay 3: 221. 1958 (Repr. ed.).

Tuberous, perennial herbs, 10–15 cm tall. Tubers 1 to 2, globose-ellipsoid. Leaves 2 or rarely 3–4, somewhat coriaceous, lying flat on the ground, distichous with sheathing base, very variable, 2–10 × 1.8–8 cm, oblong-obovate to suborbicular, acute or obtuse at apex, margins hyaline, 8–12-nerved. Flowers white, pedicellate, bracteate in 5–20 cm long, 2–many-flowered lax or dense racemes. Bracts 1–3 × 0.5–1.5 cm long, acute, 5–7-nerved. Sepals unequal, white; dorsal sepal 0.5–1 × 0.3–0.5 cm, ovate, concave, mucronulate at tip, 3-nerved. Lateral sepals 0.7–1.5 × 0.3–0.5 cm, much broader than dorsal sepal, ovate, oblique at base, 7-nerved. Petals white, 2-partite; upper segment 0.5–0.9 × 0.3–0.5 cm, subfalcate, oblique towards base, acute at tip, 3-nerved; lower segment filiform. Lip white, 3-partite, 1–2 cm long; thick, spurred; lobes subequal, spreading; lateral lobes acute slightly longer than midlobe. Spur 2–3 cm long, clavate, longer than ovary. Ovary 1.5–2 cm long, curved. Capsule 2–2.8 cm long, ribbed.

Illus.: Sant. and Kapad. *op. cit.*

Fls. and Frts.: Aug.–Dec. *Exsiccata*: MMS 1376, MPB 5102

Occurrence: Common.

Distribution: Ahmednagar, Akola, Amaravati, Aurangabad, Kolhapur, Mumbai, Nashik, Pune, Raigad, Ratnagiri, Sangali, Satara, Sindhudurg and Thane districts of Maharashtra.

Ecological Note: Species distributed throughout, grow in grasses and with scented flowers.

Taxonomic Note: Small herbs. Leaves few, radical, fleshy, flat on ground. Flowers white, faintly scented. Anthers distinctly tricuspidate at tip.

11. **Habenaria heyneana** Lindl., Gen. Sp. Orchid. 320. 1835; Hook. f. Fl. Brit. India 6: 148. 1890; Cooke, Fl. Pres. Bombay 3: 225. 1958 (Repr. ed.); Santapau and Kapad. Orch. Bombay 32, t. 5, f. 15–16. 1966; Lakshmi. in Sharma *et al.* (ed.) Fl. Maharashtra St. Monocotyledones 38. 1996. *H. subpubens* A. Rich. in Ann. Sci. Nat. 2, 15: 75, t, 4C. 1841; Hook. f *op. cit.*; Cooke, *op. cit.*

Tuberous, perennial herbs, 10–50 cm tall. Tubers 1 or 2, ellipsoid or ovoid. Stem erect, terete with 2–3 sheaths towards base. Leaves 3–5, gradually changing into floral bracts, alternately imbricating with sheathing base, 1.5–7 × 0.5–2 cm, oblong-oblanceolate, acute at apex, margins hyaline. Flowers white, sessile or subsessile, bracteate in 2–15 cm long, 2–many-flowered dense or lax, secund spikes. Bracts 1.2–2.5 cm long, acuminate, pale green. Sepals unequal, greenish-white; dorsal sepal 0.7–1 × 0.2–0.5 cm, erect, ovate-elliptic, 1-nerved. Lateral sepals 0.8–1.2 × 0.2–0.5 cm, oblique at base, 1-nerved. Petals white, 0.5–0.8 cm, somewhat falcate, 1-nerved. Lip white, 3-lobed below the middle, 0.5–1.2 cm long; spurred; lobes unequal, spreading; lateral lobes shorter than midlobe, falcate, entire, acute; midlobe longer than lateral lobes, oblong. Spur white, 0.6–1 cm long, clavate, shorter than ovary. Ovary 1–4 cm long, sessile. Capsule 1.2–2 cm long, ribbed.

Illus.: Sant. and Kapad. *op. cit.*; Abraham and Vatsala Introduction to Orchids f. 232. 1981.

Fls. and Frts.: Aug.–Dec. *Exsiccata*: MMS 274, MPB 5311

Occurrence: Common.

Distribution: Ahmednagar, Kolhapur, Nashik, Pune, Raigad, Ratnagiri, Sangali, Satara and Sindhudurg districts of Maharashtra.

Ecological Note: Grow at high altitudes amidst grasses and on plateaus.

Taxonomic Note: Flowers secund on one side and with large midlobe than side lobes.

12. **Habenaria hollandiana** Sant. Fl. Purandhar 126. 1958; Sant. and Kapad. Orch. Bombay 40. 1966; Lakshmi. in Sharma *et al.* (ed.) Fl. Maharashtra St. Monocotyledones 39. 1996. *H. affinis* Wight, Ic. t. 1707. 1852 non D. Don, 1825; Hook. f. Fl. Brit. India 6: 149. 1890; Cooke, Fl. Pres. Bombay 3: 225. 1958 (Repr. ed.).

Tuberous, perennial herbs, 20–40 cm tall. Tubers 2, ellipsoid or ovoid. Stem erect, terete with 2–3 sheaths towards base. Leaves 3–5, 2–12 × 0.5–3.5 cm, oblong-oblanceolate, clustered little above the base of stem, gradually changing into floral bracts, alternately imbricating with sheathing base, acute at apex, margins hyaline. Flowers green, subsessile, bracteate in 15–25 cm long, many flowered dense racemes. Bracts 1–1.5 cm long, acuminate, as long as or shorter than the ovary. Sepals green, unequal; dorsal sepal 3–3.5 × 1.5–2 mm, erect, concave, ovate-oblong, 3-nerved. Lateral sepals 2.5–3 × 1.5–2 mm, oblique at base, 3-nerved. Petals green, 3–3.5 mm, oblong-lanceolate. Lip green, longer than sepals, 3-lobed, almost to the base; spurred; lobes unequal, spreading; lateral lobes longer than midlobe, falcate, entire, acute; midlobe shorter and stouter than lateral lobes, oblong, straight. Spur 0.6–0.8 cm long, clavate, shorter than ovary. Capsule 1–1.4 cm long, fusiform, ribbed.

Illus.: Wight, *op. cit.*

Fls. and Frts.: Oct.–Dec. *Exsiccata*: s.n. Bhogaonkar

Occurrence: Very rare.

Distribution: Amaravati and Pune districts of Maharashtra.

Ecological Note: Grow amidst grasses. This species is collected and reported by Dr. (Mrs.) Prabha Bhogaonkar, Head, Department of Botany, Government Vidarbha Mahavidyalaya, Amravati and Vinod Devarkar from Semadoh of Amaravati District.

Taxonomic Note: Leaves radical, clustered little above the base. Spur shorter than ovary. Flowers green; midlobe of lip straight; spur incurved.

13. **Habenaria longicorniculata** Grah., Cat. Bombay Pl. 202. 1839; Sant. and Kapad. Orch. Bombay 29, t. 1, f. 1-1. 1966; Lakshmi. in Sharma *et al.* (ed.) Fl. Maharashtra St. Monocotyledones 39. 1996. *H. longecalcarata* A. Rich. Ann. Sci. Nat. 2, 15: 71, t. 3B. 1841; Hook. f. Fl. Brit. India 6: 141. 1890; Cooke, Fl. Pres. Bombay 3: 223. 1958 (Repr. ed.).

Tuberous, perennial usually tall herbs, upto 120 cm. Tubers 2, unequal, ellipsoid or ovoid. Stem erect, terete with 2–3 sheaths towards base. Leaves 3–10, 3–23 × 1.5–3.5 cm, oblong-oblanceolate, pentastichously clustered little above the base of stem, gradually changing into floral bracts, alternately imbricating with sheathing base, acute at apex, margins hyaline. Flowers 1–6, white, subsessile, bracteate in 25–85 cm long, lax racemes. Bracts many, unequal, 3–6.5 cm long, acuminate. Sepals subequal, white, 5–7-nerved; dorsal sepal 9–12 × 5–7 mm, erect, concave, ovate-oblong, 3-nerved. Lateral sepals 1–1.5 × 0.5–0.7 mm, much oblique at base, 3-nerved. Petals white, subfalcate, spathulate, 3-nerved, 1–1.3 × 0.3–0.5 cm, oblong-lanceolate. Lip white 1.5–3 cm, much longer than sepals, clawed, 3-lobed; spurred; lobes unequal, spreading; lateral lobes as long as or longer than midlobe, falcate, inwardly directed, serrate; midlobe shorter and stouter than lateral lobes, linear-oblong, straight. Spur 15–20 cm long, subclavate, much longer than ovary. Capsule 2.5–3 cm long, fusiform, ribbed.

Illus.: Sant. and Kapad. *op. cit.*; Abraham and Vatsala Introduction to Orchids f. 239. 1981.

Fls. and Frts.: Aug.–Dec. *Exsiccata*: MMS 2040, MPB 5201

Occurrence: Very common.

Distribution: Ahmednagar, Akola, Amaravati, Aurangabad, Kolhapur, Mumbai, Nashik, Pune, Raigad, Ratnagiri, Sangali, Satara, Sindhudurg and Thane districts of Maharashtra.

Ecological Note: Species widely spread. Grow in open areas amidst grasses. With much longer spur which contain sweet nectar. Plants in eastern tehsil flowers first and at higher altitudes flowers late.

Taxonomic Note: Leaves clustered at the base. Lateral lobes of lip broader than midlobe. Spur 2 or 3 times longer than ovary. Ovary long pedicelled.

14. Habenaria marginata Coleb. var. ***marginata*** Coleb. in Hook. Exot. Fl. 2, 17: 136. 1824; Hook. f. Fl. Brit. India 6: 150. 1890; Cooke, Fl. Pres. Bombay 3: 226. 1958 (Repr. ed.); Sant. and Kapad. Orch. Bombay 33, t. 5, f. 17. 1966; Lakshmi. in Sharma *et al.* (ed.) Fl. Maharashtra St. Monocotyledones 40. 1996.

Tuberous, perennial, upto 35 cm tall. Tubers 1–2, unequal, ellipsoid or ovoid. Stem erect, terete with 2–5 sheaths towards base. Leaves 2–7, 3–10 × 1.5–3.5 cm, oblong-lanceolate or elliptic-lanceolate, clustered at the base of stem, gradually changing into floral bracts, alternately imbricating with sheathing base, acute or obtuse at apex, margins yellowish. Flowers numerous, yellow or greenish-yellow, sessile, bracteate in 20–25 cm long, dense racemes. Bracts many, unequal, 1.5–3.5 cm long, acute, 1-nerved. Sepals unequal, green; dorsal sepal 8–12 × 5–7 mm, erect, concave, suborbicular, 5-nerved. Lateral sepals 5–7 × 2–3 mm, much oblique at base, 5-nerved, drooping. Petals yellow, falcate, spathulate, 3-nerved, 0.5–0.7 × 0.3–0.5 cm, oblong-lanceolate. Lip yellow 1–1.3 cm long, longer than sepals, clawed, 3-lobed; spurred; lobes unequal; lateral lobes as long as or longer than midlobe, straight, spreading, inwardly directed hook shaped at apex; midlobe shorter and stouter than lateral lobes, linear-oblong, straight, clavate at the apex. Spur yellowish, 1–1.2 cm long, subclavate, as long as ovary. Capsule 1.2–1.5 cm long, fusiform, ribbed.

Illus.: Sant. and Kapad. *op. cit.*

Fls. and Frts.: Aug.–Dec. *Exsiccata*: MMS 177, MPB 5200

Occurrence: Very common.

Distribution: Ahmednagar, Amaravati, Aurangabad, Beed, Bhandara, Chandrapur, Gadchiroli, Gondiya, Hingoli, Jalgaon, Kolhapur, Mumbai, Nagpur, Nanded, Nashik, Osmanabad, Parbhani, Pune, Raigad, Ratnagiri, Sangali, Satara, Sindhudurg, Thane and Vardha districts of Maharashtra.

Ecological Note: Grow in muddy places near bushes.

Taxonomic Note: Leaves radical, clustered at the base. Flowers yellow, numerous and dense. Floral bracts equal to or longer than ovary. Lateral lobes of lip straight, spreading, inwardly directed hook shaped at apex.

15. Habenaria marginata Coleb. var. **flavescens** (Hook. f.) T. Cooke, Fl. Pres. Bombay 2: 721. 1907 [3: 226. 1958 (Repr. ed.)]. Lakshmi. in Sharma *et al.* (ed.) Fl. Maharashtra St. Monocotyledones 41. 1996. *H. flavescens* Hook. f. Fl. Brit. India 6: 150. 1890. *H. marginata* f. var. *flavescens* Blatt. et McC. in J. Bombay nat. Hist. Soc. 36: 24. 1932; Sant. and Kapad. Orch. Bombay 34. 1966.

Tuberous, perennial, upto 35 cm tall. Tubers 1–2, unequal, ellipsoid or ovoid. Stem erect, terete with 2–5 sheaths towards base. Leaves 2–7, 3–10 × 1.5–3.5 cm, oblong-lanceolate or elliptic-lanceolate, scattered all over the stem, gradually changing into floral bracts, alternately imbricating with sheathing base, acute or obtuse at apex, margins yellowish. Flowers few in lax racemes, yellow or greenish-yellow, sessile, bracteate. Bracts many, unequal, 1.5–3.5 cm long, acute, 1-nerved. Sepals unequal, green; dorsal sepal 8–12 × 5–7 mm, erect, concave, suborbicular, 5-nerved. Lateral sepals 5–7 × 2–3 mm, much oblique at base, 5-nerved, drooping. Petals yellow, falcate, spathulate, 3-nerved, 0.5–0.7 × 0.3–0.5 cm, oblong-lanceolate. Lip yellow 1–1.3 cm long, longer than sepals, clawed, 3-lobed; spurred; lobes subequal; lateral lobes as long as or longer than midlobe, straight, spreading, inwardly directed hook shaped at apex; midlobe shorter and stouter than lateral lobes, linear-oblong, straight, clavate at the apex. Spur yellowish, 1–1.2 cm long, subclavate, as long as ovary; ovary distinctly beaked. Capsule 1.2–1.5 cm long, fusiform, ribbed.

Illus.: Lakshmi. in Sharma *et al.*, *op. cit.* 40.

Fls. and Frts.: Aug.–Dec. *Exsiccata*: MMS 169

Occurrence: Very rare.

Distribution: Kolhapur, Ratnagiri and Sindhudurg districts of Maharashtra.

Ecological Note: Grow in wet grasslands.

Taxonomic Note: Leaves scattered on the base. Flowers yellow, few and lax. Lateral lobes straight, spreading, inwardly directed hook shaped at apex.

16. Habenaria multicaudata Sedgew. in Rec. Bot. Surv. India 6: 352. 1919; Blatt. and McC. in J. Bombay nat. Hist. Soc. 36: 16. 1932; Sant. and Kapad. Orch. Bombay 14, t. 2, f. 6. 1966; Lakshmi. in Sharma *et al.* (ed.) Fl. Maharashtra St. Monocotyledones 41. 1996.

Tuberous, perennial slender herbs, upto 30 cm tall. Tubers 2, unequal, ellipsoid. Stem erect, terete with 3–7 sheaths towards base. Leaves 3–7, 5–20 × 1.2–5 cm, elliptic-oblong, pentastichously clustered in the middle of stem, gradually changing into floral bracts, alternately imbricating with sheathing base, narrowed at base, acute or acuminate at apex, margins hyaline. Flowers 2–10, green, pedicellate, bracteate in 10–15 cm long, lax racemes. Bracts many, unequal, 1.5–1.7 cm long, acuminate, shorter than ovary. Sepals unequal, green, 3-nerved; dorsal sepal 5–7 × 2–3 mm, erect, concave, ovate-oblong, 3-nerved. Lateral sepals 0.7–1 × 0.4–0.6 cm, much oblique at base, 3-nerved. Petals green, bipartite; upper segment, linear-filiform, subfalcate, 0.5–0.7 cm long. Lower segment 1.4–1.8 cm long, randomly contorted, filiform. Lip green, 2.5–3 cm, much longer than sepals, 3-lobed; spurred; lobes unequal, spreading; lateral lobes much longer than midlobe, upto 2.5 cm long, randomly contorted; midlobe 1.5 cm long, shorter than lateral lobes, linear-filiform, straight, randomly contorted. Spur dark green, 1–1.5 cm long, subclavate; ovary beaked. Capsule green, 2.5–3 cm long, fusiform, ribbed.

Illus.: Sant. and Kapad. *op. cit.*; Abraham and Vatsala Introduction to Orchids f. 215. 1981.

Fls. and Frts.: Aug.–Oct. *Exsiccata*: MMS 197

Occurrence: Rare.

Distribution: Akola, Kolhapur and Satara districts of Maharashtra.

Ecological Note: Grow on denuded places in moist, thick and shady places in evergreen and semi-evergreen forests at higher altitudes between 800–100 meters above mean sea level. Flowers little scented in evening.

Taxonomic Note: Leaves clustered at the middle of stem. Flowers green. Petals two partite. Sepals without filiform tips; lobes of lip randomly contorted.

17. **Habenaria ovalifolia** Wt., Ic.t. 1708. 1851; Hook.f. Fl. Brit. India 6: 149. 1890; Sant. and Kapad. Orch. Bombay 37, t. 5, f. 18. 1966; Lakshmi. in Sharma *et al.* (ed.) Fl. Maharashtra St. Monocotyledones 41. 1996.

Tuberous, perennial herbs, usually upto 90 cm tall. Tubers 2, unequal, ellipsoid. Stem erect, terete with 3–5 sheaths towards base. Leaves 3–7, 5–20 × 1.5–7 cm, oblong-oblanceolate, pentastichously clustered little above the middle of stem, alternately imbricating with sheathing base, acute at apex, margins hyaline, crimped. Flowers many, green or yellowish-green, subsessile, bracteate in lax 12–45 cm long racemes. Bracts many, unequal, 3–6.5 cm long, acuminate. Sepals subequal, green, 1-nerved; dorsal sepal 4–7 × 2–3 mm, erect, concave, ovate-oblong, 1-nerved. Lateral sepals 4–6 × 2–4 mm, much oblique at base, 1-nerved. Petals green, subfalcate, spathulate, 1-nerved, 5–6 × 3–5 mm, obliquely ovate, obtuse, 1-nerved. Lip green, fleshy, 0.7–1.2 cm, longer than sepals, clawed, 3-lobed; spurred; lobes unequal, spreading; lateral lobes 0.5–0.7 × 0.2–0.3 cm, spreading, acute at apex; midlobe shorter and stouter than lateral lobes, linear-oblong, 3-nerved, longitudinally grooved on the lower surface, inflexed, retuse or subretuse at apex. Spur 15–20 cm long, green, subclavate, much longer than ovary. Capsule 1.5–2 cm long, fusiform, ribbed.

Illus.: Sant. and Kapad. *op. cit.*

Fls. and Frts.: Aug.–Dec. *Exsiccata*: MMS 1460

Occurrence: Rare.

Distribution: Ahmednagar, Kolhapur, Nashik, Pune, Ratnagiri and Sindhudurg districts of Maharashtra.

Ecological Note: Grow on open places amidst grasses.

Taxonomic Note: Leaves clustered about middle of stem. Flowers green. Lateral lobes of lip not broader than midlobe; midlobe of lip inflexed; spur straight.

18. **Habenaria panchganensis** Sant. and Kapad. in J. Bombay nat. Hist. Soc. 54: 478. 1957 and Orch. Bombay 27, t. 6, f. 24. 1966; Lakshmi. in Sharma *et al.* (ed.) Fl. Maharashtra St. Monocotyledones 42. 1996.

Tuberous, perennial upto 25 cm tall herbs. Tubers 1–2, unequal, ellipsoid or ovoid. Stem erect, terete with 2–3 sheaths towards base. Leaves 3–7, radical, 2–10 × 0.5–2.5 cm, oblong-oblanceolate, alternate, distichously clustered at the base of stem, alternately imbricating with sheathing base, acute at apex, margins papillate. Flowers white, pedicellate, bracteate, few to many in secund or subsecund, lax or dense 5–20 cm long racemes. Bracts 1–2, unequal, 1–1.3 cm long, acuminate. Sepals unequal, white, 3-nerved; dorsal sepal 0.9–1.2 × 0.5–0.7 cm, erect, concave, ovate-oblong, 3-nerved. Lateral sepals 1–1.5 × 0.5–0.7 cm, much oblique at base, subfalcate, 3-nerved. Petals white, 1–1.3 × 0.3–0.5 cm, obliquely ovate, orbicular at the base and acuminate at the apex, 3-nerved. Lip white 1.5–2.5 cm long, much longer than sepals, 3-lobed little above the middle; spurred; lobes unequal, spreading; lateral lobes 0.6–1 cm broad, as long as or shorter than midlobe, falcate, inwardly directed, crenulate or denticulate, 5–7-nerved; midlobe longer than lateral lobes, broadly or narrowly lanceolate, entire, straight, 5-nerved. Spur 0.8–1.5 cm long, subclavate, much longer than ovary. Capsule 2.5–3 cm long, fusiform, ribbed.

Illus.: Sant. and Kapad. *op. cit.*

Fls. and Frts.: Aug.–Dec. *Exsiccata*: MMS 167, MPB 5143

Occurrence: Frequent.

Distribution: Ahmednagar, Kolhapur, Ratnagiri, Sangali, Satara and Sindhudurg districts of Maharashtra.

Ecological Note: Grow on plateaus along with *Eriocaulon, Utricularia, Smithia* and grasses. Flowers showy and fragrant.

Taxonomic Note: Leaves radical, distichously clustered at base. Flowers secund or subsecund. Lip more than 1.5 times longer than that of lateral sepals; spur much shorter than ovary.

19. Habenaria perrottetiana A. Rich., Ann. Sci. Nat. (Series 2) 15: 74, t. 4B, 1841; Hook. f. Fl. Brit. India 6: 164. 1890; Fischer in Gamble, Fl. Pres. Madras 3: 1471. 1957 (Repr. ed.)

Tuberous, perennial usually upto 60 cm tall herbs. Tubers 1–2, unequal, ellipsoid or ovoid. Stem erect, terete with 4–7 sheaths. Leaves 3–7, 3.5–8 × 1.5–2.5 cm, ovate-lanceolate, scattered on stem, gradually changing into floral bracts, alternately imbricating with sheathing base, acute at apex, margins hyaline. Flowers yellow, 1–6, subsessile, bracteate in upto 20 cm long, lax racemes. Bracts many, unequal, foliaceous, acuminate. Sepals subequal, greenish-yellow; dorsal sepal 9–12 × 5–7 mm, erect, concave, ovate-oblong, 3-nerved. Lateral sepals 1–1.5 × 0.5–0.7 mm, much oblique at base, 3-nerved. Petals yellow, subfalcate, spathulate, 3-nerved, 1–1.3 × 0.3–0.5 cm, oblong-lanceolate. Lip yellow 1.5–3 cm, much longer than sepals, clawed, 3-lobed at base; spurred; lobes subequal; lateral lobes as long as or little longer than midlobe, falcate, inwardly directed, serrate; midlobe shorter and stouter than lateral lobes, linear-oblong, straight. Spur 15–20 cm long, subclavate, much longer than ovary. Capsule 2.5–3 cm long, fusiform, ribbed.

Illus.: Abraham and Vatsala Introduction to Orchids f. 237. 1981.

Fls. and Frts.: Aug.–Dec. *Exsiccata*: MMS 4341

Occurrence: Very rare.

Distribution: Kolhapur district of Maharashtra.

Ecological Note: Grow amidst grasses at higher altitudes.

Taxonomic Note: Leaves scattered along stem. Flowers 1–6, yellow, secund in one or two rows.

20. Habenaria plantaginea Lindl., Gen. Sp. Orchid. 323. 1835; Hook. f. Fl. Brit. India 6: 141. 1890; Cooke, Fl. Pres. Bombay 3: 224. 1958 (Repr. ed.); Sant. and Kapad. Orch. Bombay 28, t. 6, f. 25. 1966;

Lakshmi. in Sharma *et al.* (ed.) Fl. Maharashtra St. Monocotyledones 42. 1996.

Tuberous, perennial upto 50 cm tall herbs. Tubers 1–2, unequal, ellipsoid or ovoid. Stem erect, terete with 2–3 sheaths towards base. Leaves 3–7, 2–20 × 0.5–3.5 cm, oblong-lanceolate, pentastichously clustered at the base of stem, gradually changing into floral bracts, alternately imbricating with sheathing base, acute at apex, margins papillate. Flowers white, subsessile, bracteate, few to many in secund or subsecund, upto 30 cm long racemes. Bracts many, unequal, 3–6.5 cm long, acuminate, 3-nerved. Sepals unequal, white, 3-nerved; dorsal sepal 4–7 × 2.5–5 mm, erect, concave, ovate, 3-nerved, minutely denticulate. Lateral sepals 4–5 × 2–4 mm, much oblique at base, 3-nerved, minutely denticulate. Petals 4–6 × 1.3–2 mm, white, subfalcate, spathulate, 1-nerved, oblong-lanceolate, minutely denticulate. Lip white 7–12 × 7–9 mm, much longer than sepals, clawed, 3-lobed, about little below the middle, spurred; lobes unequal, spreading; lateral lobes as long as or little longer than midlobe, falcate, entire or minutely serrate; midlobe shorter than lateral lobes, linear-oblong, straight, subacute. Spur white, 1.5–4 cm long, clavate, much longer than ovary. Capsule 2–4 cm long, fusiform, ribbed.

Illus.: Sant. and Kapad. *op. cit.*

Fls. and Frts.: Aug.–Dec. *Exsiccata*: MMS199, MPB 5946

Occurrence: Rare.

Distribution: Ahmednagar, Amaravati, Bhandara, Chandrapur, Kolhapur, Nashik, Pune, Raigad, Ratnagiri, Sangali and Satara districts of Maharashtra.

Ecological Note: Grow on hill slopes near bushes.

Taxonomic Note: Leaves radical. Lateral lobes of lip broader than midlobe. Spur 2 or 3 times longer than ovary.

21. **Habenaria rariflora** A. Rich. in Ann. Sci. Nat. 2, 15: 70, t. 2D. 1841; Hook. f. Fl. Brit. India 6: 136. 1890; Cooke, Fl. Pres. Bombay 3: 221. 1958 (Repr. ed.); Sant. and Kapad. Orch. Bombay 15, t. 3, f. 9–10. 1966; Lakshmi. in Sharma *et al.* (ed.) Fl. Maharashtra St. Monocotyledones 43. 1996.

Tuberous, perennial terrestrial, or rarely lithophytic herbs, usually upto 30 cm tall. Tubers 1–2, unequal, ellipsoid or ovoid. Stem erect or pendulous, terete with 2–3 close sheaths at the base. Leaves 2–5, 3–15 × 0.5–5 cm, linear-lanceolate to oblong-oblanceolate, radical, clustered at the base of stem, alternately imbricating with sheathing base, acute at apex, margins hyaline. Flowers 1–7, white, pedicellate, bracteate in 25–85 cm long, slender secund or subsecund, ebracteate lax racemes. Bracts white, unequal, 3-nerved, acuminate. Sepals unequal, white, 3–5-nerved; dorsal sepal 6–12 × 5–7 mm, erect, concave, ovate, 5-nerved. Lateral sepals 8–15 × 5–7 mm, much oblique at base, 5-nerved. Petals white, subfalcate, spathulate, 3-nerved, 1–1.3 × 0.3–0.5 cm, oblong-lanceolate. Lip white 1.5–3 cm, much longer than sepals, clawed, 3-lobed; spurred; lobes unequal, spreading; lateral lobes as long as or longer than midlobe, falcate, inwardly directed, serrate; midlobe shorter and stouter than lateral lobes, linear-oblong, straight. Spur 15–20 cm long, subclavate, much longer than ovary. Capsule 2.5–3 cm long, fusiform, ribbed.

Illus.: Sant. and Kapad. *op. cit.*; Abraham and Vatsala. Introduction to Orchids f. 217. 1981.

Fls. and Frts.: Aug.–Dec. *Exsiccata*: MMS 178, 202, MPB 5443

Occurrence: Frequent.

Distribution: Ahmednagar, Kolhapur, Nashik, Pune, Raigad, Ratnagiri, Sangali, Satara, Sindhudurg and Thane districts of Maharashtra.

Ecological Note: Grow in loose soil on steep slopes, dripping rocks along with *Tripogon spp.*, *Utricularia striatula*, *Impatiens acaulis* and grasses.

Taxonomic Note: Spur longer than flower.

22. **Habenaria roxburghii** Nicols. in Sald. and Nicols. Fl. Hassan Dt. 834. 1976; Lakshmi. in Sharma *et al.* (ed.) Fl. Maharashtra St. Monocotyledones 43. 1996. *H. platyphylla* Spr. Syst. Veg. 3: 690. 1826; Hook. f. Fl. Brit. India 6: 140. 1890; Cooke, Fl. Pres. Bombay 3: 222. 1958 (Repr. ed.); Sant. and Kapad. Orch. Bombay 19. 1966.

Tuberous, perennial upto 40 cm tall herbs. Tubers 1–2, unequal, ellipsoid or ovoid. Stem erect, terete with 2–5 sheaths towards base. Leaves 2–4, 2–12 × 1.5–8 cm, oblong-elliptic to orbicular, fleshy, clustered at the base of stem, lying flat on the ground, margins papillate. Flowers white, pedicellate, bracteate, few to many in dense or lax 15–25 cm long racemes. Bracts many, unequal, 1.5–2 cm long, acuminate. Sepals unequal, white, 3-nerved; dorsal sepal 6–10 × 5–7 mm, erect, concave, ovate-oblong, 3-nerved. Lateral sepals 8–11 × 0.5–0.7 mm, much oblique at base, 3-nerved. Petals white, subfalcate, linear-spathulate, 3-nerved, 1–1.3 × 0.3–0.5 cm, oblong-lanceolate. Lip white, 5–8 mm long, clawed, 3-lobed; spurred; lobes unequal, spreading; lateral lobes 3-4 mm long, falcate, 1–2-nerved; midlobe 4 mm long, linear-oblong, straight, 3-nerved. Spur 2.5–3.5 cm long, subclavate, much longer than ovary. Capsule 2–2.5 cm long, fusiform, ribbed.

Illus.: Anandrao, T. Conservation of Wild Orchids of Kodagu District in the Western Ghats. f. 150. 1998.

Fls. and Frts.: Aug.–Dec.

Occurrence: Very rare.

Distribution: Amaravati, Kolhapur, Latur, Nanded, Parbhani and Yeotmal districts of Maharashtra.

Ecological Note: It grows amidst bushes and rock crevices in scrub forests.

Taxonomic Note: Leaves ovate-orbicular, flat on ground. Flowers numerous, white in dense racemes. Lip shorter than lateral sepals, three lobed; spur twice as long as the ovary.

23. Habenaria stenopetala Lindl., Gen. SP. Orchid. 319. 1835; Hook. f. Fl. Brit. India 6: 134. 1890; Cooke, Fl. Pres. Bombay 3: 219. 1958 (Repr. ed.); Sant. and Kapad. Orch. Bombay 9. 1966; Lakshmi. in Sharma *et al.* (ed.) Fl. Maharashtra St. Monocotyledones 44. 1996.

Tuberous, perennial herbs. Tubers 1–2, unequal, ellipsoid or ovoid. Stem erect, terete with 3–5 sheaths at the base. Leaves 3–10, 3–15 × 1–4.5 cm, oblong to lanceolate, pentastichously clustered about the middle of stem, gradually changing into floral bracts, alternately imbricating with sheathing base, acute or acuminate at apex, margins wavy, hyaline. Flowers many, brownish-green, pedicellate, bracteate in 15–25 cm long dense racemes. Bracts many, unequal, 2.5–3 cm long, acuminate. Sepals subequal, ovate, entire, 3-nerved; dorsal sepal 8–11 × 4–7 mm, erect, concave, ovate-oblong, 3-nerved, with 2-4 mm long, straight or curved acumen. Lateral sepals 1-1.5 × 0.5–0.7 mm, much oblique at base, 3-nerved, with 2–4 mm long, straight or curved acumen. Petals white, 2-partite; upper segment 7–9 mm long, erect linear, 3-nerved; lower segment minute, 2–3 mm long, lanceolate, acute. Lip greenish-white, 3-lobed almost to the base, spurred; lobes unequal, spreading; lateral lobes 5–6 mm long, shorter than midlobe; midlobe 1–12 mm long, much longer than lateral lobes, linear-oblong, curved downwards, acuminate. Spur 1.8–2.5 cm long, subclavate, longer than ovary. Capsule 1.2–1.5 cm long, fusiform, ribbed.

Illus.: Blatt. and McC. in J. Bombay nat. Hist. Soc. 36: 14, t. 1. 1932.

Fls. and Frts.: Aug.–Dec. *Exsiccata*: *s.n.* Kulkarni

Occurrence: Very rare.

Distribution: Amaravati, Kolhapur, Mumbai, Pune and Thane districts of Maharashtra.

Ecological Note: Grow in grasslands on hill slopes.

Taxonomic Note: Leaves pentastichously clustered about the middle of stem. Flowers many, brownish-green, in dense racemes. Petals two partite. Sepals with filiform tips; lateral lobes .of lip less than half as long as midlobe.

24. **Habenaria suaveolens** Dalz. in Hook. J. Bot. and Kew Gard. Misc. 2: 263. 1850; Hook. f. Fl. Brit. India 6: 140. 1890; Cooke, Fl. Pres. Bombay 3: 222. 1958 (Repr. ed.); Sant. and Kapad. Orch. Bombay 27. 1966; Lakshmi. in Sharma *et al.* (ed.) Fl. Maharashtra 1: 44. 1996.

Tuberous, perennial herbs. Tubers 1–2, unequal, ellipsoid or ovoid. Stem erect, terete with 2–3 sheaths towards base. Leaves 5, 1.5–2 × 1–3.5 cm, oblong, clustered at the base of stem spreading on the ground, acute at apex. Flowers small, white, bracteate in 25–30 cm long dense spikes. Bracts many, unequal, 1.2–2 × 0.4–0.7 cm, oblong or oblong-lanceolate, acuminate, as long as or shorter than ovary. Sepals subequal, white, acute; dorsal sepal ovate. Lateral sepals ovate, much oblique at base. Petals white, subfalcate, ovate, acute. Lip white as long as or slightly longer than sepals, 3-lobed; spurred; lobes unequal, spreading; lateral lobes broader and slightly shorter than midlobe, serrate; midlobe longer than lateral lobes, linear, obliquely truncate, denticulate at apex. Spur pendulous, filiform, scarcely clavate, as long as or shorter than ovary. Capsule 1–1.3 cm long, fusiform, ribbed.

Fls. and Frts.: Aug.–Sept.

Occurrence: Very rare.

Distribution: Sindhudurg district of Maharashtra.

Ecological Note: Grow amidst grasses in moist deciduous forests.

Taxonomic Note: Leaves clustered at the base of stem spreading on the ground. Lip as long as or shorter than the lateral sepals. Lateral lobes of lip broader than midlobe. Spur shorter than or equaling ovary.

25. **Habenaria viridiflora** (Sw.) R. Br. Prodr. 312. 1810; Hook. f. Fl. Brit. India 6: 150.1890; Sant. and Kapad. Orch. Bombay 35, t. 5, f. 19. 1966; Lakshmi. in Sharma *et al.* (ed.) Fl. Maharashtra 1: 44. 1996. *Orchis viridiflora* Sw. in Act. Holm. 706. 1800. *Habenaria viridiflora* var *dalzellii* Hook. f. *op. cit.*; Cooke, Fl. Pres. Bombay 3: 226. 1958 (Repr. ed.).

Tuberous, perennial herbs. Tubers 1–2, unequal, ellipsoid or ovoid. Stem erect, terete with 2–3 sheaths towards base. Leaves 3–5, 3–8 × 0.5–1.5 cm, linear-oblong, radical, clustered at the base of stem, alternately imbricating with sheathing base, mucronate at apex, margins hyaline. Flowers white, subsessile, bracteate in 8–15 cm long, few to many flowered lax racemes. Bracts many, unequal, 7–20 × 4–6 mm long, narrowly lanceolate, acuminate, much shorter than ovary. Sepals unequal, obtuse, 3-nerved; dorsal sepal 2–3 × 1.5–2 mm, erect, concave, ovate or suborbicular, 3-nerved. Lateral sepals 2–2.5 × 1.5–1.7 mm, oblique at base, 3-nerved. Petals 2–2.5 × 1.3–1.5 mm, subfalcate, 3-nerved. Lip 3.5–4.5 mm, 3-partite at the base; spurred; lobes subequal, spreading, variable in length, linear, acute; midlobe as long as or little shorter than lateral lobes, linear-oblong, acute. Spur 1–1.5 cm long, subclavate, longer than ovary. Capsule 1–1.5 cm long, fusiform, ribbed.

Illus.: Sant. and Kapad. *op. cit.*

Fls. and Frts.: Aug.–Sept. *Exsiccata*: MMS 4245

Occurrence: Very rare.

Distribution: Kolhapur and Sindhudurg districts of Maharashtra.

Ecological Note: Grow amidst grasses.

Taxonomic Note: Leaves radical, clustered at the base. Flowers white. Floral bracts up to half as long as ovary.

Excluded Species

1. **Habenaria caranjensis** Dalz. in Hook. J. Bot. and Kew Gard. Misc. 2: 262. 1850 et in Dalz. and Gibs. Bombay Fl. 267. 1861; Hook. f. Fl. Brit. India 6: 166. 1890; Cooke, Fl. Pres. Bombay 3: 228. 1958 (Repr. ed.); Sant. and Kapad. Orch. Bombay 41. 1966; Lakshmi. in Sharma *et al.* (ed.) Fl. Maharashtra St. Monocotyledones 45. 1996.

Lower leaves somewhat rounded, upper oblong-lanceolate, 3-nerved. Bracts acuminate, shorter than ovary. Flowers small, yellow. Upper sepals rounded; petals ovate, obtuse. Lip 3-partite; midlobe oblong, obtuse; lateral ones shorter, cuneate, truncate at apex. Spur clavate, shorter than ovary.

Fls. and Frts.: Not known.

Occurrence: Very rare.

Distribution: Mumbai (Bombay) district of Maharashtra.

Ecological Note: Grow amidst grasses at lower altitudes in moist and dry deciduous forests.

Taxonomic Note: Leaves flat on ground. Flowers white or greenish-white. Lip longer than lateral sepals, three partite; lobes subequal or lateral ones longer segments of lip.

Note: The above description of the species is given by Dalzell in 'Bombay Flora' of Dalzell and Gibson. There are no specimens in Herbarium, Kew and plant does not appear to have been collected by any subsequent collector after other than Dalzell. Santapau and Kapadia stated that the original locality given by Dalzell is the islands of 'Carunja near Bombay'. But there is no place with such a name near Bombay. There is, however, a place called Karanja or Caranja on the sea coast across the harbour of Bombay, which seems to be the locality of Dalzell's collection.

Results and Discussion

Habenaria Willd., mainly the tuberous genus is represented by about 800 species distributed throughout the world especially tropics and temperate regions of the world (Sathish Kumar and Manilal, 1994). In India it is represented by 72 species, of which 38 species are endemic to their geographical areas. Of the 72 species recorded for India, 43 species occur in Peninsular India and rest are restricted to Himalayas and Andmans. About 30 species of *Habenaria* are recorded to be endemic to Peninsular India. Present survey of orchids of Maharashtra resulted in collection of one more species from the State increasing the number of species to 22 species and three varieties.

The species endemic to Peninsular India viz. *Habenaria crassifolia*, *H. foliosa* var. *foetida*, *H. foliosa* var. *foliosa*, *H. foliosa* var. *gibsonii*, *H. grandifloriformis*, *H. heyneana*, *H. longicorniculata*, *H. multicaudata*, *H. ovalifolia*, *H. panchganensis*, *H. perrottetiana*, *H. rariflora*, *H. roxburghii*, and *H. suaveolens* are also found in Maharashtra. *Habenaria panchganensis* and *H. suaveolens* are exclusively endemic to Maharashtra State.

Field observations on distribution of Orchidaceae in Maharashtra have revealed that the species have their own ecological preferences. *Habenaria marginata* grows in wet grasslands. *Habenaria crinifera* grow as lithophytic on tree trunks. Some species like *Habenaria crinifera* and *H. rariflora* grow as lithophyte on hanging rocks.

Terrestrial species of *Habenaria* with their different ecological affinities towards the substratum have varied habitat preferences. *Habenaria diphylla* and *H. suaveolens* prefer to grow in sandy gravely soils on the lateritic plateaus in Konkan region. *Habenaria heyeneana*, *H. panchganiensis* prefer to grow on plateaus of Sahyadri ranges.

Habenaria crassifolia, *H. digitata*, *H. foliosa* var. *foliosa*, *H. foliosa* var.

gibsonii, H. furcifera, H. multicaudata, H. ovalifolia and *H. plantaginea* prefer to grow amidst on hills slopes at higher altitudes.

Habenaria diphylla, H. marginata var. *flavescens* and *H. suaveolens* are restricted to dry deciduous forests of coastal plains of Konkan of the State, while *Habenaria crassifolia, H. crinifera, H. digitata, H. foliosa* var. *foetida, H. foliosa* var. *foliosa, H. foliosa* var. *gibsonii, H. furcifera, H. heyneana, H. longicorniculata, H. multicaudata, H. ovalifolia, H. panchganensis, H. plantaginea, H. rariflora* and *H. viridiflora* are distributed in dry and moist deciduous forests of Western Ghats of the State. *Habenaria roxburghii* is restricted to dry deciduous and scrub forests of eastern region of Maharashtra. *Habenaria foliosa* var. *foetida, H. grandifloriformis* and *H. marginata* var. *marginata* are distributed throughout the State of Maharashtra.

The flowers in *Habenaria commelinifolia, H. crinifera, H. foliosa* var. *foliosa, H. foliosa* var. *gibsonii, H. grandifloriformis, H. panchganensis, H. plantaginea, H. rariflora, H. roxburghii* and *H. viridiflora* are white. While in *Habenaria crassifolia, H. digitata, H. diphylla, H. foliosa* var. *foetida, H. foliosa* var. *gibsonii, H. furcifera, H. multicaudata, H. ovalifolia, H. stenopetala* and *H. suaveolens* the flowers are green and *Habenaria marginata* var. *flavescens, H. marginata* var. *marginata* and *H. perrottetiana* are with yellow flowers.

Present studies indicated that careful collections and critical observations would reveal new discoveries and better understanding of *Habenaria* species of our country.

Acknowledgement

Authors are thankful to Head, Department of Botany, Shivaji University, Kolhapur for providing research facilities and to Department of Science and Technology, Delhi for financial assistance. Authors are also thankful to Dr. (Mrs.) Prabha Bhogaonkar, Head, Department of Botany, Government Vidarbha Mahavidyalaya, Amravati for encouragement and providing necessary help.

REFERENCES

Almeida, S.M. 1990. *Flora of Sawantwadi, Maharashtra, India J. Econ. Tax. Bot. Addl. Ser.*, 2 Vols.

Blatter, E. and McCann, C. 1931–32. Revision of the flora of the Presidency of Bombay. *J. Bombay nat. Hist. Soc.* **32:** 14–33, 281–298, 408–435, 622–649; **33:** 7–25, 229–243, 480–496, 753–775; **34:** 12–26; **35:** 13–31, 254–275, 484–495, 722–736; **36:** 13–28, 307–320, 524–537, 781–795, **37:** 15–35, 255–277, 532–548, 764–779; **38:** 6–18.

Cooke, T. 1901–08. *The Flora of the Presidency of Bombay*. London. (B.S.I. Reprint). Calcutta, Vols. I–III, 1958.

Dalzell, N.A. and Gibson, A. 1861. *The Bombay Flora*. Bombay.

Deshpande, S., Sharma, B.D. and Nayar, M.P. 1993–95. *Flora of Mahabaleshwar and it's adjoinings, Maharashtra*. BSI. Vol. I and II.

Diwakar, P.G. and Sharma, B.D. 1999. *Flora of Buldhana District, Maharashtra State*. BSI Calcutta.

Gammie, G.A. 1905–1912. The Orchids of Bombay Presidency. *J. Bombay nat. Hist. Soc.* **16:** 429–433, 562–569; **17:** 31–37, 940–942; **18:** 88–91, 586–590, 833–834; **19:** 139–141, 624–626; **20:** 126–129, 597–602; **21:** 171–174, 1129–1130.

Gibson, A. 1863. *A handbook to the forest flora of Bombay presidency*. Bombay.

Graham, J. 1839. *A catalogue of the plants growing in Bombay and it's vicinity, spontaneous cultivated or introduced as far as they have been ascertained*. Bombay.

Hooker, J.D. 1872–1897. *The Flora of British India*. London. Vols. 1–7.

Kamble, S.Y. and Pradhan, S.G. 1988. *Flora of Akola district, Maharashtra*. BSI.

Karthikeyan, S. and Ananadkumar. 1993. *Flora of Yevatmal District, Maharashtra. BSI. Fl. India ser.* 3. pp. XXXII + 344, colour pl. 4, fig. 27.

Karthikeyan, S., Jain. S.K., Nayar, M.P. and Sanjappa. 1989. *Florae Indicae Enumeratio: Monocotyledonae*. BSI.

Kothari, M.J. and Moorthy, S. 1993. *Flora of Raigad District. Maharashtra State* BSI.

Kulkarni, B.G. 1988. *Flora of Sindhudurg. BSI. Fl. India ser.* 3, pp. XX + 605, pl. 11, figs. 6.

Lakshminarsimhan, P. and Sharma, B.D. 1991. *Flora of Nasik district. B.S.I. Fl. India ser.* 3, pp. 641, pl. 15, figs. 52.

Mishra, D. and Singh, N.P. 2001. *Endemic and Threatened Flowering Plants of Maharashtra*. Botanical Survey of India. Calcatta.

Naik, V.N. 1979. *The Flora of Osmanabad*. Aurangabad, Venus Publishers.

Naik, V.N. 1998. *Flora of Marathwada*, Vol 1 and 2. Amrut Prakashan, Aurangabad.

Patel, R.I. 1968. *Forest Flora of Melghat*. pp. XLVIII + 380. Bishen Singh Mahendra Pal Singh, Dehradun.

Rheede, H.A. Van. 1678–1693. *Hortus malabaricus* Vol. 1–12. Amsterdam.

Roxburgh, W. 1820 and 1824. *Flora Indica* Vol. 1–2. Serampore.

Santapau, H. 1953. *The Flora of Khandala on Western Ghats of India Rec. Bot. Surv. India* **16(1):** XXVII + 396, map, fig. 3. (Rev. ed. 1967).

Santapau, H. 1955. *Botanical collectors manual*. Govt. Press, New Delhi.

Santapau, H. 1958d. *The Flora of Purandhar or an enumeration of all phanerogamic plants discovered in Purandhar during the years 1944–56*. New Delhi. pp. 158.

Santapau, H. and Kapadia, Z. 1966. *The Orchids of Bombay*. pp. VI + 239, pl. 54. Manager of Publications, Delhi.

Sardesai, M.M. and Yadav, S.R. 2000. Family Orchidaceae in Kolhapur District. *J. Swamy Bot. Cl.* **17:** 65–68.

Sathishkumar, C. and Manilal, K.S. 1994. A Catalogue of Indian Orchids 162. p. 64 pl.

Sharma, B.D., Karthikeyan, S. and Singh, N.P. (eds.) 1996. *Flora of Maharashtra State: Monocotyledones* B.S.I., Calcutta. 1–793.

Ugemuge, N.R. 1986. *Flora of Nagpur district*. Sree Prakashan, Nagpur.

Vartak, V.D. 1966. *Enumeration of plants from Gomantak, India*. Poona.

Yadav, S.R. and Sardesai, M.M. 2002. *Flora of Kolhapur District*. Shivaji University, Kolhapur.

CHAPTER 7

THREATENED PLANTS OF ANDHRA PRADESH

T. Pullaiah

Introduction

The State of Andhra Pradesh lies between 12° 37′ and 19° 54′ North latitudes and 76° 46′ and 84° 46′ East longitudes. Geographically it occupies the middle portion of the eastern half of the Indian peninsula with an area of 2,75,068 sq kms accounting for 8.4% of the country. The state has tropical monsoon type of climate, which is generally hot and humid. The climate of the state is classified as tropical rainy type. Maximum temperature in the summer season varies between 37° and 44° and minimum temperature in the winter season varies between 14° C and 19°C. The annual rainfall in the coastal region is between 72–150 cm where as Rayalaseema region receives annual rainfall between 60–70 cms and Telangana receives annual rainfall between 75–100 cm. There are five soil types in Andhra Pradesh; these include red, black, laterite, coastal sands and skeletal soils. Vegetation types include a) Tropical semi evergreen, b) Tropical moist deciduous, c) Dry deciduous, d) Northern mixed dry deciduous type, e) Dry savanna, f) Tropical dry evergreen, g) Tropical dry evergreen scrub, h) coastal strand vegetation and i) Esturine vegetation.

* Department of Botany, Sri Krishnadevaraya University, Anantapur 515 003, A.P.

Botanical survey of India has listed 16 threatened plants in the State of Andhra Pradesh. These are listed below. Many more species are threatened and their details are also given in this article.

Valid names are given in bold letters and synonyms are given in italics.

Endangered Plants Occurring in Andhra Pradesh Listed in Red Data Book of Indian Plants.

1. **Acacia campbelli** Arn. Nova Acta Phys. Med. Acad. Caes. Lep. Casol. Nat. Cur. 18: 333. 1836.

Family: Mimosaceae

Small tree, blaze hard, yellowish white; stipular spines straight, slender, white. Leaves bipinnate, up to 2.7 cm long; pinnae 6–10 paired; leaflets minute, ovate-oblong, hairy. Flower heads pink, 8 mm across; peduncles fairly stout, bracteate about the middle. Pods papery, 5–8 cm long, obtusely pointed, straight or curved.

Godavari to Nellore, Kurnool, Cuddapah (Nayar and Sastry, 1990).

Status: Rare.

2. **Albizia thompsoni** Brandis in Indian For. 25: 284. 1899; Gamble 1: 432. 1991.

Family: Mimosaceae

Large unarmed deciduous tree, 12–13 m tall, bark black, rough, wood thick, white in colour, branches pubescent. Leaves bipinnate, 16–19 cm long, petiole short, a gland above middle of the petiole, rachis tomentose with a gland between the upper most pairs of pinnae. Pinnae 4–6 pairs, 8–16 cm long, leaflets 15–19 pairs, overlapping, oblong, slightly falcate, 1–1.5 × 0.4–0.5 cm, broadest on the lower side of the midrib, round or semicordate at base, margin entire, apex obtuse, appressed pubescent when young, ashy grey when dry. Peduncle 2–4 cm, flowers pale yellow, in globose heads, pedicel short, 3 mm; calyx 5-lobed, to 3 mm, campanulate, pubescent without, triangular, corolla 5-toothed, to 3.5 mm, pubescent without, lanceolate, acute, staminal tube 6 mm, stamens many, filaments 5 mm, monadelphous at the base, ovary 1.5 mm, sessile or shortly stalked, style to 3.5 mm, filiform. Pods moderately thick, 10–12 × 1–2.5 cm, shining, rich brown, dehiscent, seeds 6–8, brown.

Occasional in forests in Visakhapatnam and Anantapur districts. Fl. and Fr.: April–October.

Andhra Pradesh, Orissa and Tamil Nadu–endemic.

Status: Rare

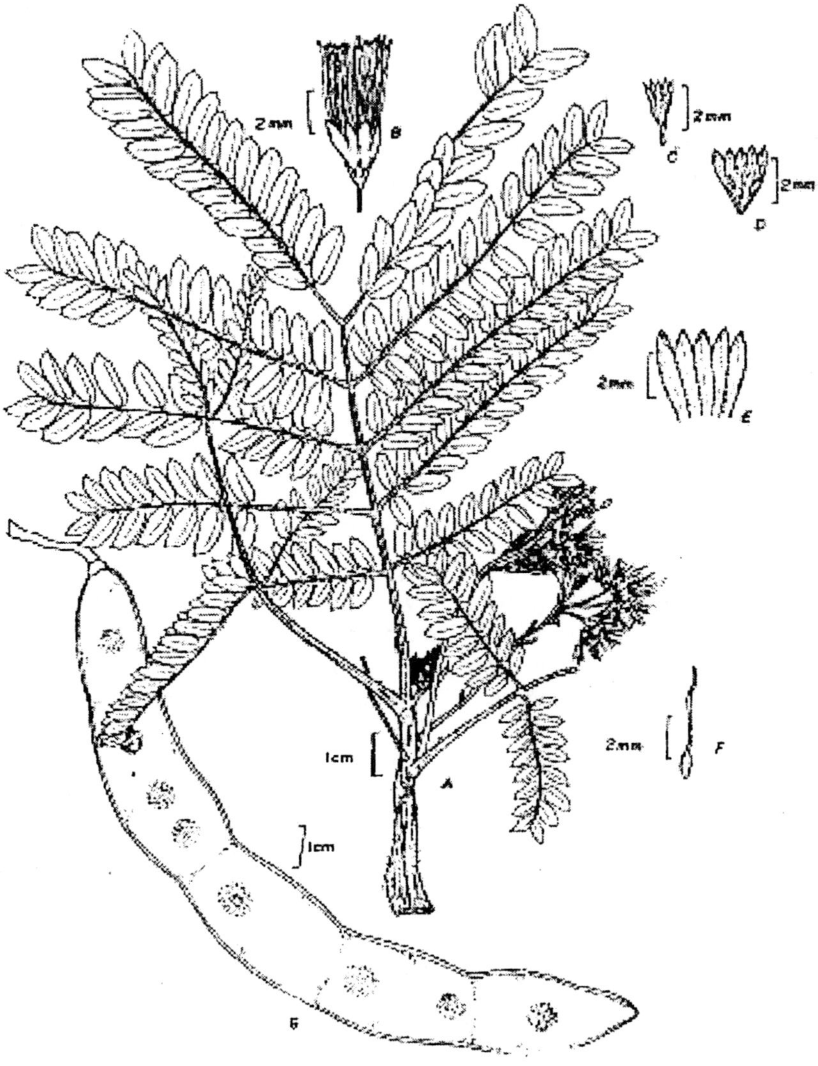

Fig.1. *Albizia thompsoni* Brandis: A.Twig, B. Flower, C. Calyx entire, D. Calyx split open, E. Corolla split open, F. Pistil, G. Pod.

3. Ceropegia spiralis Wight Ic. t. 1267. 1848; FBI 4: 66. 1883; Gamble 2: 856. 1923.

Family: Asclepiadaceae

Slender erect herb, root tuberous, tuber depressed. Occasionally twining. Stem glabrous. Leaves linear, sessile, 7–10 × 0.2–0.3 cm, chartaceous, glabrous, base acute, apex acute. Flowers axillary solitary, corolla 4–5 cm long, tube 2–2.5 cm long, subcylindric, hairy inside except at inflated base; lobes 2.0–2.5 cm long, linear, hairy within above from ovate or orbicular base, generally spirally twisted. Outer corona of 5 bifid, deltoid lobes, glabrous; inner linear, erect.

Hills of Cuddpah, Nellore and Ranga Reddy districts. It also occurs in Karnataka, Kerala and Tamil Nadu. Fl and Fr.: August–October.

This rare species was reported in Flora of Presidency of Madras from Cuddapah District based on Beddome's collection (MH 32707 – 1881). Since then it has been collected from Nellore and Ranga Reddy Districts.

Status: Vulnerable.

4. **Cycas beddomei** Dyer in Trans. Linn. Soc. Ser. II. 5: 85, t. 17. 1881.

Family: Cycadaceae

Shrub, stem 10–15 cm long. Leaves up to 1 m long; petioles ca 15 cm long, the upper one-third with few minute spines, base covered with tufted tomentum; leaflets 10–18 × 0.5 cm, linear, apex acuminate, margins strongly revolute. Male cone ca 20 × 8 cm, rusty brown, apical portion of microsporophyll of the upper half of the cone strongly deflexed, blade of megasporophyll usually narrow-lanceolate, long-acuminate; margins dentate-lobate, ovules 2 on either side. Seeds globose, ca 4 cm across.

Endemic to the Cuddapah-Tirupati range of Seshachalam range of Southern Eastern Ghats.

Status: Vulnerable.

5. **Decaschistia rufa** Craib in Kew Bull. 1912: 35. 1912; Gamble 1: 94. 1915

Family: Malvaceae

Shrub, up to 3 m tall, branches with reddish hairs and thereafter paler. Leaves almost rounded, 3.5–7.5 × 4–7 cm, base subcordate or rounded, apex shortly acuminate, acute or subobtuse, margin crenate-serrate or denticulate, both the surfaces densely stellate pubescent, more densely on the lower surface, 7-nerved at the base. Stipules linear, single or 2–3-lobed. Flowers reddish, solitary, axillary, 2.5–3 cm across. Epicalyx segments 10, connate at the base, lobes obvate-lanceolate. Calyx campanulate, lobes deltoid, fused below middle, petals

2–2.5 × 1–1.5 cm, densely stellate hairy outside, glabrous inside, staminal column 1.2 cm long, antheriferous through out, ovary ovoid, 3 × 2.2 mm. Fruit loculicidally, 10-valved, depressed globose capsule, seeds reniform, shortly pubescent.

Rare in Cuddapah and Chittoor districts.

Status: Endangered.

6. **Eriolaena lushingtonii** Dunn in Kew Bull. 1915: 88. 1915: Gamble: 109. 1915.

Family: Sterculiaceae

Trees, up to 7 m tall, bark greyish red; young branches pubescent. Leaves membranous, cordate-orbicular, 4 × 3.6 cm, stellate hairy above, white pubescent beneath, apex acute or shortly acuminate, margin shortly, irregularly dentate, base cordate, 5–7-nerved, petiole to 1.7 cm, densely stellate pubescent. Flowers yellow, axillary on 2–3-flowered cymes. Capsules ovoid, ca 1–1.5 × 0.5–0.7 cm, woody, pubescent, valves 5–8; seeds 5–8 in each locule, winged on one side top.

Rare in Nallamalais, in open slopes of moist deciduous forests between 350 m and 900 m.

Fl and Fr.: June–August.

Status: Vulnerable.

7. **Hildegardia populifolia** (Roxb.) Schott. & Endl. Melet. Bot. 33. 1832. *Sterculia populifolia* Roxb. Fl. India 3: 148. 1832; FBI 1: 361. 1874 (*populnifolia*); Gamble 1: 107. 1915.

Family: Sterculiaceae

Deciduous tree, up to 20 m tall; bark smooth, wood whitish; branchlets angular, tubercled. Leaves thin-membranous, cordate, 9.5–11.5 × 8–11 cm, glabrous, base deeply cordate, margin entire, apex long-acuminate, 7-nerved, lateral nerves 4 pairs above basal nerves, raised; petiole 5–7 cm, whitish. Flowers scarlet, polygamous in axillary and terminal, up to 15 cm long, spreading panicles. Fruit rusty brown, 2-follicled, follicle obliquely lanceolate, membranous, 10.2 cm long, base and apex 5.2 and 2.5 cm width, wings flat, inflated, strongly nerved; seeds oblong, solitary in each follicle, 1.6 × 1.2 cm, pale brown.

Fl: April, Fr.: June – Feb. Tel: Galibudagala chettu.

Ahmedullah in Nayare and Sastry (1990) noted that there is only one population comprising of about 20 trees in Kalrayan Hills. Dr. B. Ravi Prasad Rao of Sri Krishnadevaraya University surveyed the population and reports that there are atleast 40,000 trees in Southern

Andhra Pradesh and Tamil Nadu. He is of the opinion that it should be taken out of the Red List.

8. **Indigofera barberi** Gamble in Kew Bull. 1919: 222. 1919 and Fl. Pres. Madras 1: 310. 1918.

Family: Fabaceae

Erect canescent undershrub, up to 1 m tall, branchlets greyish, slender, angular, pubescent. Stipules setaceous, 4 mm. Leaves 3-foliolate, 2.5 × 1.2 cm, base cuneate, margin entire, apex obtuse, mucronate, pubescent below, glabrescent above, gland dotted,

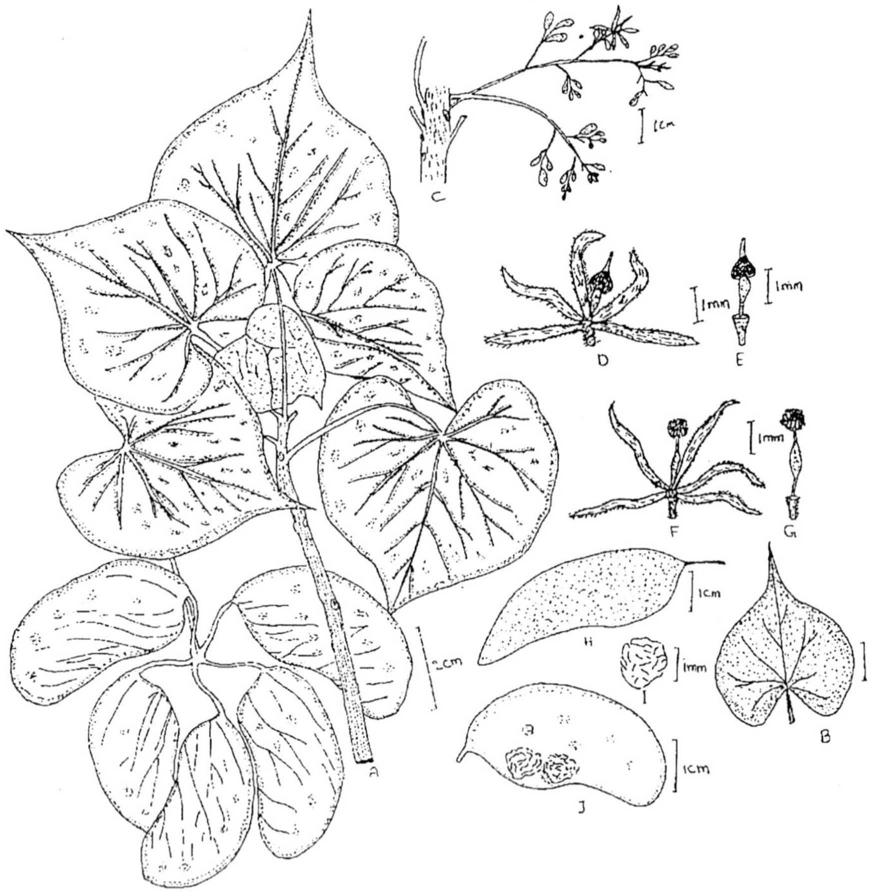

Fig. 2. *Hildegardia populifolia* (Roxb.) Schott. and Endl.: A. Twig, B. Leaf, C. Flowering twig, D. Bisexual flower, E. Pistil, F. Male flower, G. Staminal column, H. Fruit, I. Follicle open

petiole to 2.5 cm, petiolule 2 mm. Bracts linear, 3 mm. Racemes axillary, to 1.5 cm, flowers 3 mm across, pedicel 1 mm, calyx pubescent, tube 0.5 mm, lobes setaceous, to 1.5 mm, corolla red, exserted, matty tomentose without, standard oblong-obovate, 6 × 2.5 mm, wings 4.5 × 1 mm, keels 5 × 1.5 mm with 1.5 mm spur, stamens diadelphous (9+1), staminal sheath 3 mm, filaments 0.5 mm; ovary subsessile, 2 mm, style 3 mm. Pod sub-terete, 1.2 × 0.2 cm, angular, torulose, deflexed, appressed white pubescent, apex curved, seeds 4–6, oblong-ellipsoid, 2 mm.

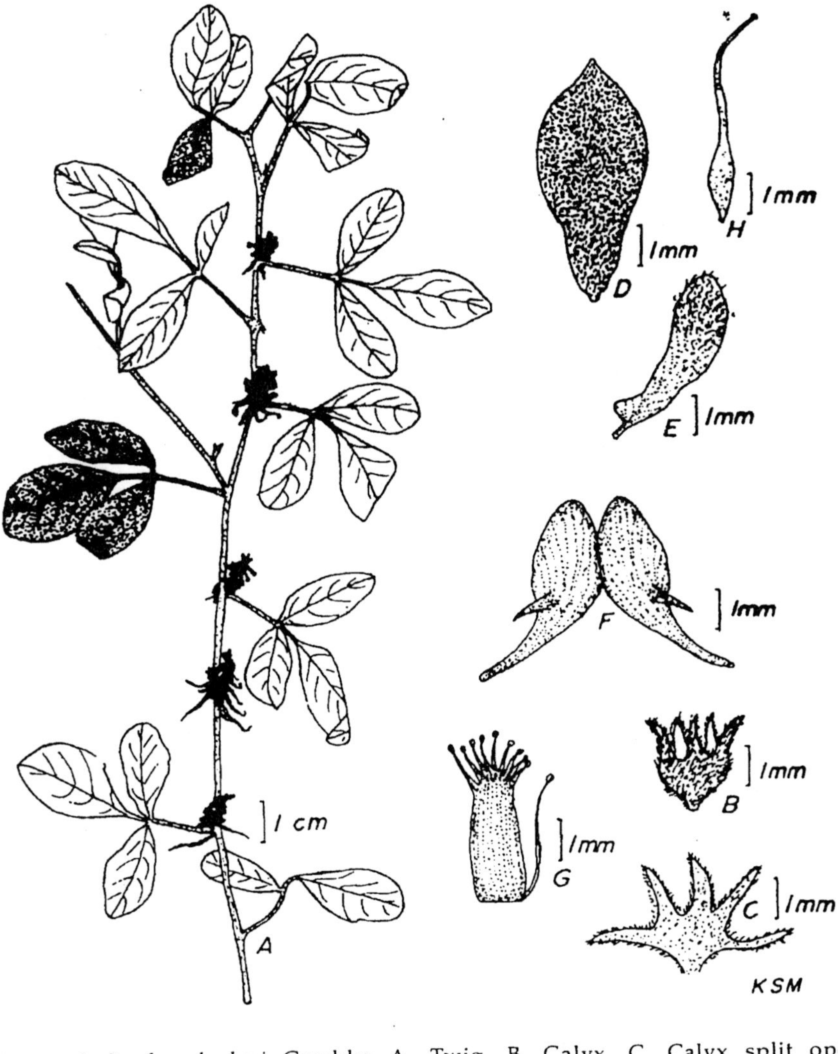

Fig. 3. *Indigofera barberi* Gamble: A. Twig, B. Calyx, C. Calyx split open, D. Standard petal, E. Wing petal, F. Keel petal, G. Staminal column, H. Pistil.

Occasional in dry deciduous forests of southern Eastern Ghats.

Fl and Fr.: September –December.

Status: Rare.

9. **Isonandra villosa** Wight Ic. t. 360. 1843; FBI 3: 538. 1882; Gamble 2: 761. 1921; *I. lanceolata* Wt. Ic. p 367. 1843.

Family: Sapotaceae

Tree, branchlets thick, obscurely hairy. Leaves coriaceous, broadly obovate, up to 7 × 6.2 cm, glabrous, apex rounded or emarginate, margin entire, base rounded; secondary nerves 12–15 pairs, regular, parallel, 5 mm apart; petiole to 3 mm. Flowers sessile in clusters on very short thick lateral spurs. Berry 1.2 × 0.5 cm, ellipsoid or oblong with a fleshy pericarp; seed 1, flattened.

Veligonda hills of Nellore.

Status: Indeterminate.

Ramaswamy collected this taxon from Veligonda hills a century ago (Fl. Pres. Madras) and C.A. Barber from Kambakkam hills in Chengalput in 1914 (10242–MH). Suranarayana and Srinivasa Rao (2002) collected from Venkatagiri Durg of Veligonda hills. Thus this species was recovered after a century.

10. **Leucas mukherjiana** Subba Rao & Kumari in Bull. Bot. Surv. India 10: 358. 1968.

Family: Lamiaceae

Decumbent herb, about 45 cm tall; stem quadrangular, covered with white, deflexed, silky villous hairs. Leaves 1.8– 3.7 × 1.4–3.4 cm, thick, ovate, base cordate, margin dentate, densely velvety on both sides. Inflorescence in whorls, axillary, few to many-flowered; bracts filiform, villous, often equaling calyx tube; calyx tube straight, 10-toothed; corolla white, bilabiate; stamens didynamous, upper pair shorter. Nutlets 4, 1 mm long.

Occasional in Cherukonda in Visakhapatnam.

Fl. and Fr.: June–January.

Status: Endangered.

11. **Nilgirianthus circarensis** (Gamble) Bremek. in Verh. Kon. Ned. Akad. Wetensch. Afd. Natuurk Tweede Sect. 41. 172. 1944. *Strobilanthes circarensis* Gamble in Kew Bull. 1923: 373. 1923 and Fl. Pres. Madras 2: 1038. 1924.

Family: Acanthaceae

An erect shrub, with sub-tetragonous, nearly glabrous stems. Leaves 5–11 × 2.5–6.2 cm, ovate, base narrowed, apex acute, main nerves 5 pairs. Flowers in short, dense hairy heads; bracts leafy, obovate; bracteoles linear, very white-silky; calyx-lobes narrowly lanceolate, scarious below, silky above; corolla pale blue. Capsule large, seeds glabrous.

As undergrowth of moist deciduous forests in Visakhapatnam district.

Fl. and Fr.: October–November.

Status: Rare.

12. **Phlebophyllum jeyporense** (Bedd.) Bremek. in Verh. Kon. Ned. Akad, Wetensch. Afd. Natuurk. Tweede Sect. 41: 169. 1944. *Strobilanthes jeyporensis* Bedd. Ic. t. 214. 1874; FBI 4: 436. 1884; Gamble 2: 1036. 1924.

Family: Acanthaceae

Large erect shrub, up to 3 m tall, branches lenticellate, ribbed, Leaves submembranous, up to 17.5 × 10 cm, opposite, ovate, base rounded and decurrent on the petiole, margin crenulate, apex acuminate. Young leaves glabrous above and softly pubescent beneath, glabrous or hairy beneath when mature. Petioles up to 6 cm long. Flowers white with blue tips in axillary, panicled spikes. Bracts up to 1.4 × 0.3 cm, acuminate, recurved in fruit, glandular hairy without; bracteoles up to 0.8 cm long, linear-lanceolate, glandular hairy without; calyx up to 0.8 cm long, densely glandular hairy without, tube up to 0.4 cm long; lobes 5, subequal, linear; corolla up to 1.7 cm long, straight, lobes 5, unequal, spreading; stamens 2, adnate to the base of the corolla tube, filaments up to 0.9 cm long, hairy in the lower half; ovary up to 0.1 cm long, linear-oblong, seated on a 0.1 cm long disk; style up to 1 cm long, thickened and slightly hairy at top. Fruit capsule, up to 1.5 cm long, clavate, beaked, ciliate at apex. Seeds 2, ovate-oblong, densely hairy except on the areolae.

In Visakhapatnam and East Godavari districts.

Status: Endangered.

13. **Phyllanthus narayanaswamii** Gamble in Kew Bull. 329. 1925 and in Fl. Pres. Madras 2: 1289. 1925.

Family: Euphorbiaceae

Undershrub, up to 60 cm tall, monoecious; branchlets many, from a stout rootstock, branchlets terete, winged. Leaves 0.6–1.2 × 0.3–0.6

cm, elliptic, obliquely at base, entire, thickened, recurved, rounded-obtuse, sometimes slightly emarginate or mucronate; petioles short; stipules peltate, lacerate, acuminate, hastate at base. Flowers axillary, solitary; pedicels of male flowers to 5 mm, female flowers to 5.5 mm. Capsule 4 mm across, 3-lobed, verrucose glandular; seeds to 2 mm, golden brown, covered with red glands.

Rare in the hills of East Godavari and Visakhapatnam districts.

Fl and Fr.: July–October.

Status: Endangered.

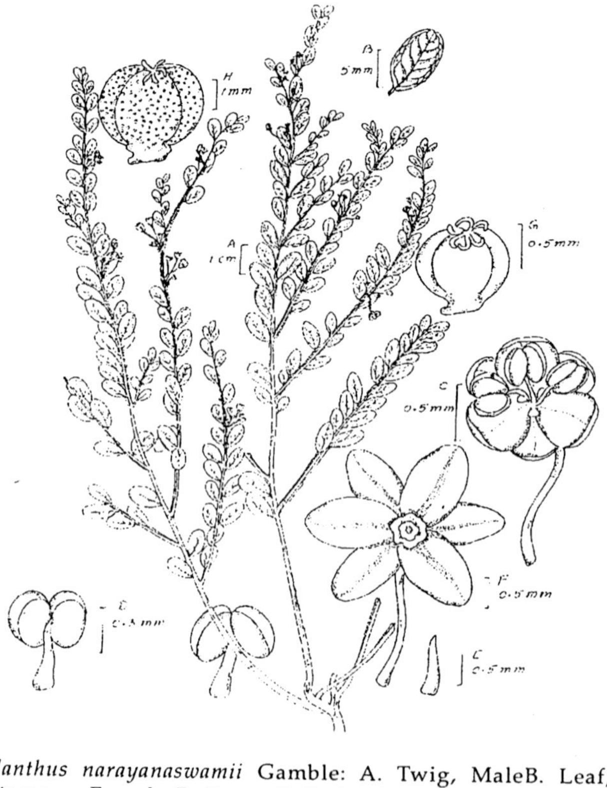

Fig. 4. *Phyllanthus narayanaswamii* Gamble: A. Twig, MaleB. Leaf, C. Flower, D. Stamens. Female E. Bract, F. Perianth, G. Pistil, H. Capsule.

14. **Pimpinella tirupatiensis** Balakr. & Subr. in Bull. Bot. Surv. India 2: 427–428. 1960.

Family: Apiaceae

Erect woody herb, up to 55 cm tall, completely glabrous, stem simple below, branched above, terete, striate, branches alternate, root tuberous. Basal and cauline leaves simple, base deeply cordate, apex acute-obtuse or acute, 1.7–3.8 × 1.3–3.5 cm, sheathing at base, upper cauline leaves palmately 3-partite, segments oblanceolate to linear-lanceolate, acute, narrowed towards base, cut and sometimes lobed, uppermost leaves cut into small setaceous lobes or often reduced to mere sheaths. Flowers white in compound umbels, primary umbels with 3–6 slender rays, bracts 0–1, partial umbels 5-16-flowered, bracteoles 1–2. Cremocarp 1.5 mm long, ovoid, papillose, scabrous.

Endemic to Tirumala hills in Chittor district.

Fl and Fr.: September–January.

Status: Endangered.

15. Rhynchosia beddomei Baker in Hook.f. Fl. Brit. India 2: 222. 1876; Gamble 1: 374. 1918.

Family: Fabaceae

Erect shrub, up to 1.5 m tall, branchlets terete, white tomentose. Leaves 3-foliolate, to 8 cm, leaflets ovate lanceolate, 2–6 × 1.5–2.5 cm, subcoriaceous, base obtuse, apex acute, margin entire, velvety, white pubescent on both surfaces, strongly nerved beneath, petiole to 2 cm, petiolule to 3 mm. Racemes short, axillary or terminal, flowers 8 mm across; calyx 5 mm long, lobes oblong, obtuse, longer than the corolla, the lowest shortest; corolla yellow. Pod suborbicular, 8 × 5 mm, thinly canescent, seed 1, strophiolate.

Rare, in dry deciduous and moist deciduous forests of Southern Eastern Ghats in Anantapur, Chittoor and Cuddapah districts.

Fl.: November–January; Fr.: February–April.

Andhra Pradesh, Tamil Nadu and Karnataka — endemic.

Status: Rare.

16. Toxocarpus longistigma (Roxb.) Wight & Arn. ex Steud. Nomencl. Bot. 2: 694. 1841.

Family: Asclepiadaceae

Climbing shrubs, stems glabrous, young shoots rusty pubescent. Leaves ovate-elliptic, up to 10.5 × 7 cm, base rounded or acute, margin entire and wavy, apex abruptly acuminate, glabrous on both sides, glandular at base of midrib on upper side; petioles up to 1.7 cm long, sparsely rusty pubescent, grooved, twisted. Flowers yellow, small, in rusty pubescent dichotomous lax cymes; cymes shorter than leaves; peduncles interpetiolar; up to 1 cm long, rusty pubescent; bracts linear, acute, up to 0.2 cm long, rusty pubescent; pedicels up to 0.1 cm long,

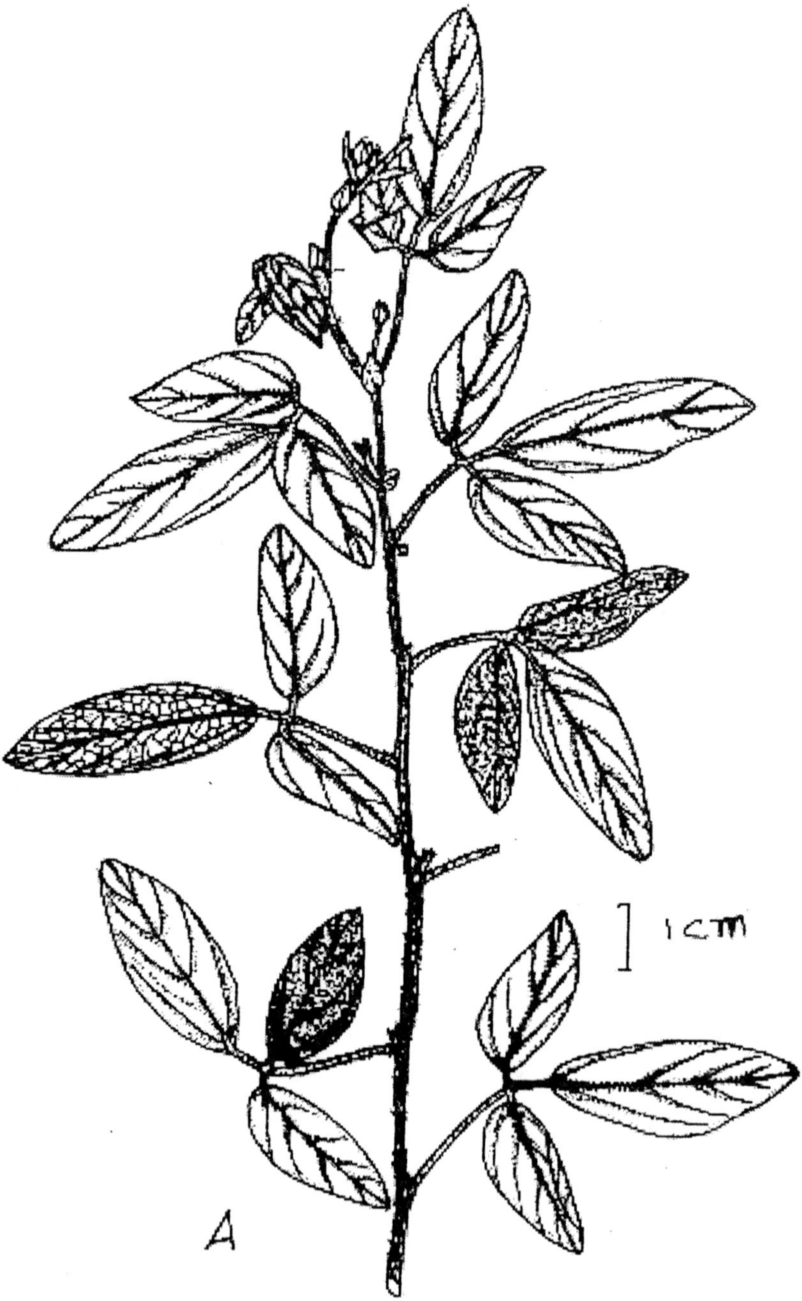

Fig. 5. *Rhynchosia beddomei* Baker: A. Twig.

terete, rusty pubescent; calyx rusty pubescent without, divided almost to the base; lobes each up to 0.3 × 0.1 cm, elliptical, obtuse at apex; corolla up to 0.4 cm long, glabrous without, minutely hairy at throat within, lobes contorted in bud, each up to 0.2 × 0.05 cm, lanceolate, obtuse; corona up to 0.15 cm long, style up to 0.2 cm long. Follicles 2, divaricate, up to 1.2 cm broad towards base, tapering to a point towards apex, rusty pubescent while young, sparsely rusty pubescent when mature. Seeds flat, elliptical, chocolate brown, up to 1 × 0.3 cm; coma silky, white, up to 2.7 cm long.

Rare in East Godavari and Visakhapatnam districts (Nayar and Sastry, 1987).

Status: Endangered.

Threatened Plants to be Added to the Red Data Book of Indian Plants

17. **Aerides maculosum** Lindl. Bot. Reg. t.58. 1845; Fischer 3: 1442. 1928. *Saccolobium speciosum* Wight Ic. t. 1674-75. 1851.

Family: Orchidaceae

Epiphytic herb, stems elongate, branching. Leaves leathery, up to 22 × 5 cm, fleshy, channeled. Inflorescence longer than the leaves, pendulous and branched. Flowers large, pale-rose coloured, fragrant.

Recorded only from Kurnool district by Rama Rao *et al.* Also occurs in Maharashtra, Orissa.

Status: Vulnerable.

18. **Andrographis beddomei** C.B. Clarke in Hook.f. Fl. Brit. India 4: 506. 1884; Gamble 1054. 1924.

Family: Acanthaceae

A much branched nearly glabrous undershrub. Leaves 3.5–7.5 × 2.2–3.7 cm, elliptic or obovate, subsessile, glabrous. Flowers white with purple spots on the lower lip in few-flowered short racemes, calyx lobes glandular pubescent. Stamens 2, anthers bearded. Capsule 1.5 cm long, 8-seeded.

Rare in Simhachalam hills in Visakhapatnam district and Guvvalacheruvu forest in Cuddapah district.

Fl. and Fr.: October–March.

Status: Not evaluated.

19. **Aphyllorchis montana** (Thw.) Reichnb.f. in Linnaea 41: 57. 1876; FBI 6: 116. 1890; Fischer 3: 1457. 1928. *Apaturia montana* Thw. Enum. Pl. Zeyl. 301. 1861.

Family: Orchidaceae

A terrestrial leafless saprophyte, stem sheathed, sheaths near the base of the stem tubular, loose and close together. Flowers straw-coloured with purple tips in racemes. Capsule oblong-pyriform.

Occasional in hills of Chittoor district.

Fl. and Fr.: February–August.

It also occurs in Himalayas.

Status: Rare.

20. **Argyreia pilosa** Arn. Pug. Pl. Ind. Or. 38. 1836; FBI 4: 189. 1883; Gamble 2: 908. 1923.

Family: Convolvulaceae

A hirsute climbing shrub. Leaves deltoid-ovate, base truncate, margin entire, apex acute, sub-coriaceous, hirsute above, strigose below, petiolate. Flowers blue, in corymbose cymes. Fruit berry.

Occasional in the hills of Adilabad and Visakhapatnam districts.

Fl. and Fr.: July–November.

Status: Endangered.

21. **Cajanus cajanifolius** (Haines) van der Maesen in Agric. Univ. Wageningen Pap. 85. 4. 1985. *Atylosia cajanifolia* Haines in J. Asiat. Soc. Beng. N.s. 15: 312. 1920.

Family: Fabaceae

A large erect shrub, up to 4 m tall. Leaves trifoliolate, leaflets oblong-lanceolate, base and apex acute, margin entire, pale green above and silky beneath, lateral leaflets smaller. Flowers red in axillary racemes. Pod velvety, 2–5-jointed, constrictions oblique, seeds dark gray.

Occasional in Srikakulam, Visakhapatnam and East Godavari districts.

Status: Endangered.

22. **Cyathocline manilaliana** Prabhakar Raju et Venkata Raju in Rheedea 9: 151.1999.

Family: Asteraceae

Erect aromatic herb, up to 40 cm tall, stem simple, often branched from base, ultimate branches in dichotomies, pubescent with hairs and glands, axillary buds densely covered by persistent silky hairs. Leaves simple, radical and cauline, cauline ones alternate, sessile, puberulous and glandular on both surfaces, coriaceous, uppermost ovate, lower spathulate or oblong, 1–13 × 0–4.3 cm, lyrately or deeply lobed, lobes triangular, dentate, acute. Heads few, in compound corymbs, terminal, pink purple, heterogamous, not rayed, involucral bracts 2-seriate, subacuminate, margins not scarious, anthers 2–3.5 mm long, anther base deeply sagittate with sharp auricles. Achenes of female florets oblong or elliptic, occasionally narrowly rostrate at both ends, glabrous, of disc minute, around 0.2 mm, sterile.

Rare in Pochera fields in Adilabad district.

Fl.: January–March, Fr.: February–April.

Status: Endangered.

23. Dicliptera beddomei C.B.Clarke in Hook.f. Fl. Brit. India 4: 551. 1885; Gamble 2: 1073. 1924.

Family: Acanthaceae

A slender herb, branches striate. Leaves up to 14 × 8 cm, ovate, minutely pubescent, base acuminate. Flowers in trichotomous cymes on lax panicles, the branches of the panicles white-hairy; bracts minutely pubescent, villous without, stamens 2. Capsules 8 mm long, clavate, pubescent; seeds verrucose.

Rare in Nallamalais. Collected from type locality after a century by D.Ali Moulali. Fl and Fr.: September–February.

Status: Not Evaluated.

24. Dimorphocalyx kurnoolensis Venkataraju & Pullaiah in Bot. Bull. Acd. Sin. 35: 201. 1994.

Family: Euphorbiaceae

Small tree, to 3 m tall, dioecious; branchlets terete, striate, pubescent. Leaves 5–19 × 2–6 cm, elliptic-lanceolate or obovate, attenuate at base, stipules deltoid. Male flowers in cymes; peduncles pubescent, bracts lanceolate, densely pubescent, acute-acuminate. Flowers 4–5 mm across, subsessile; tepals in two whorls, 5 + 5; outer (sepals) cupular, 5-lobed, connate, adpressed-pilose without, lobes suborbicular, inner (petals) free, oblong, obtuse, often emarginate; stamens biseriate, 5 + 11–17; outer basally connate to the inner staminal column; filaments 1.8 mm; anthers 1.2 mm across, widely oblong; inner monadelphous, staminal column 7 mm. Anther 0.8 mm across, orbicular, connectives 0.5 mm long, acute; disc of glands 5, free, ovate-oblate, hairy at top. Female flowers few, in short pedunculate racemes; flowers 5–8 mm across; pedicel 2 mm, pubescent, tepals in two whorls,

Fig. 6. *Dicliptera beddomei* C.B.Clarke: A. Twig, B. Inflorescence, C. Inflorescence bract, D. Flower bract, E. Bracteole, F. Calyx, G. Corolla, H. Stamen, I. Pistil, J. Capsule.

5 + 5, outer (sepals) cupular, shortly 5-lobed, connate, adpressed-pilose without, lobes suborbicular, 2 × 2.5 mm; inner (petals) free, oblong-obtuse, often emarginate; ovary 4 × 3 mm, adpressed-pilose; styles 3, connate at base, each 2-fid, from above the middle, papillose; disc of glands as in male flower. Capsule 1–1.3 × 1.3–1.7 cm, subglobose, depressed, adpressed-pilose, 3-lobed, deeply furrowed. Fruiting calyx (sepals) deeply 5-lobed, lobes 5.5 × 3.5 mm, adpressed pilose without.

Owk forest of Kurnool district. About 500–1000 plants are growing here.

Fl. and Fr.: January–March. Vern. Name: Tel.: Erra Tandra.

Status: Endangered.

25. Eulopia ochreata Lindl. in Journ. Linn. Sic. 3: 24. 1858; FBI 6: 2. 1890; Fischer 3: 1435. 1928.

Family: Orchidaceae

A terrestrial glabrous herb with fleshy tuberous roots. Leaves 3–5, elliptic-lanceolate. Flowers in terminal racemes.

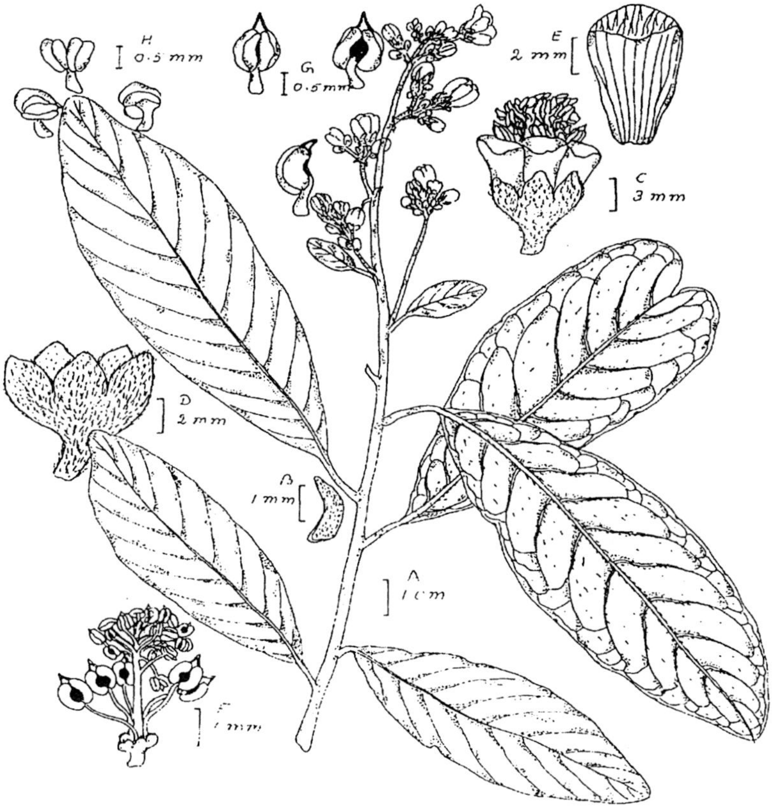

Fig. 7. *Dimorphocalyx kurnoolensis* Venkataraju and Pullaiah – Male: A. Twig, B. Bract, C. Flower, D. Outer tapals (sepals), E. Inner tepals (petals), F. Androecium, G and H. Anthers.

In hills of Visakhapatnam (A.W. Lushington, cf. Fischer).

Status: Vulnerable.

26. Euphorbia senguptae Balakr. & Subr. in Bull. Bot. Surv. India 2: 175. 1960. *Chamaesyce senguptae* (Balakr. & Subr.) Raju & Rao in Phytologia 37: 454. 1977.

Family: Euphorbiaceae

Erect herb, up to 20 cm tall, dichotomously branched; branchlets pubescent near nodes. Leaves 0.7–1.2 × 0.3–0.7 cm, elliptic- oblong, obliquely truncate, subcordate at base, serrulate, acute; petiole short, to 1 mm; stipules 0.5 mm, ovoid. Cyathia solitary, axillary; involucre 1.8 × 1.5 mm, campanulate, lobes laciniate; glands 4, equal, small,

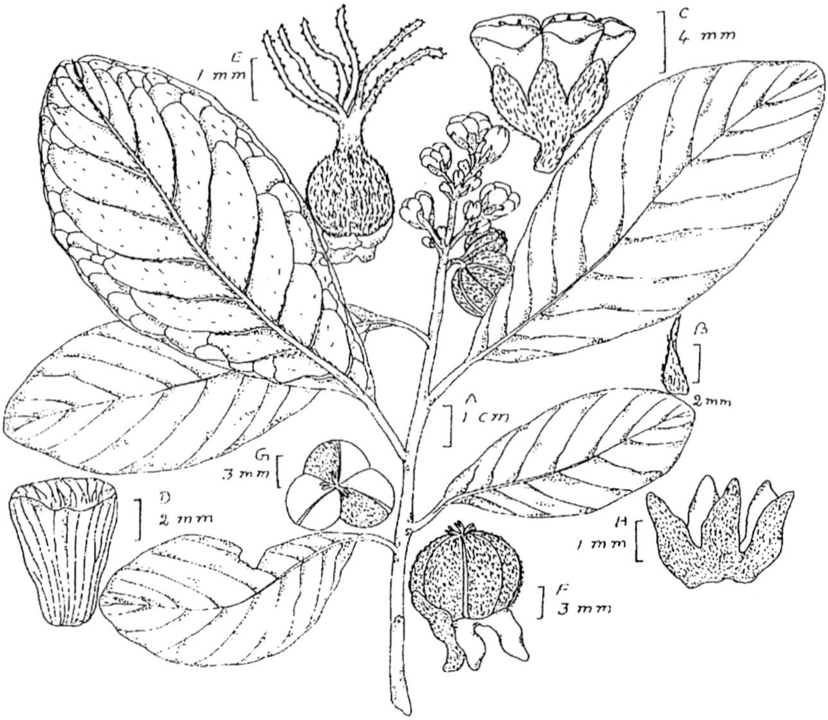

Fig. 8. *Dimorphocalyx kurnoolensis* Venkataraju & Pullaiah – Female: A. Twig, B. Bract, C. Flower, D. Inner tapal (petal), E. Pistil, F. Capsule with calyx, G. Capsule upper view), H. Fruiting calyx.

partially oblong; appendages laciniate. Male florets 8–12. Female floret laterally pendulous, capitate. Capsule 2 mm; seeds 1.2 mm, 4-angled.

Occasional in forests of Anantapur and Kurnool districts.

Fl. and Fr.: September – December.

Status: Vulnerable.

27. Habenaria rariflora A. Rich. Ann. Sci. Nat. 2: 15: 70 t. 2 D. 1841; FBI 6: 136; 1890; Fischer 1469. 1928.

Family: Orchidaceae

A terrestrial erect herb, up to 23 cm tall. Leaves usually 3 or more, linear to linear-oblong, rarely ovate, narrowed to the sheath, 2–10 × 0.5–2.5 cm, sometimes complicate. Flowers white, 1–3 in short racemes; bracts ovate, acuminate, pedicels erect, about as long as the ovary; lateral sepal ovate-oblong, acute, subfalcate; dorsal sepal broadly ovate, subacute, petals 2-partite, upper segments ovate, acuminate, oblique,

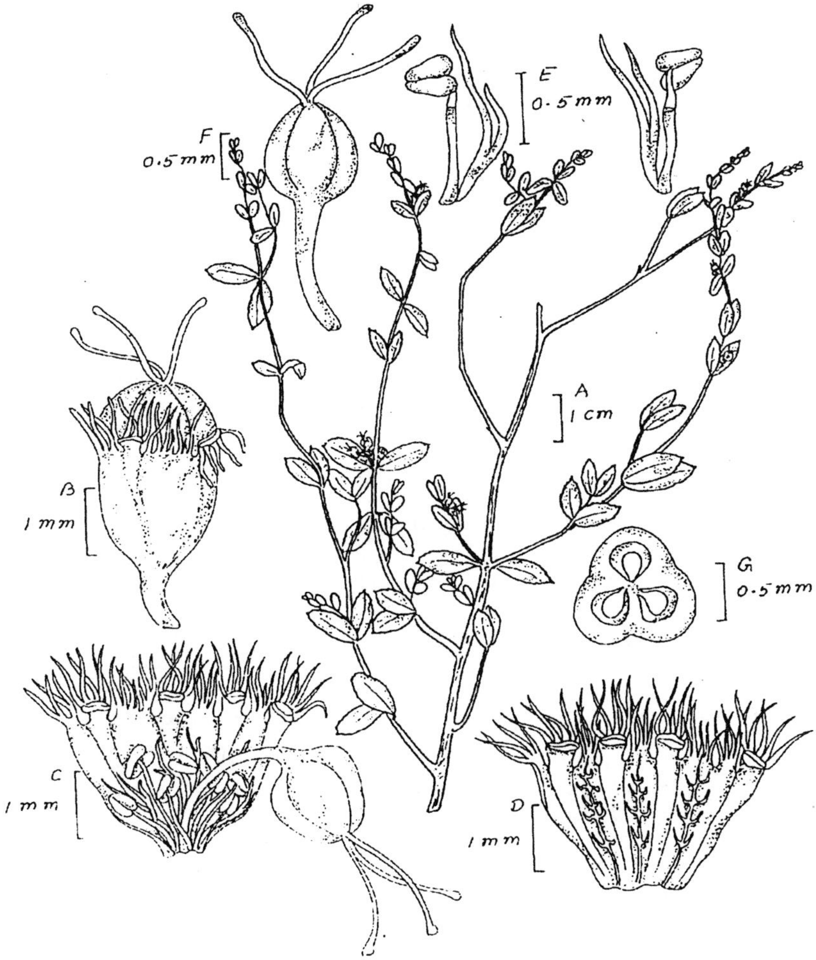

Fig. 9. *Euphorbia senguptae* Balakr. & Subr.: A. Twig, B. Cyathium, C. Cyathium split open, D. Involucre split open, E. Stamens with bracteoles, F. Pistil, G. Ovary T.s.

lower segments longer than the upper, linear, acute, lip 3-partite; lateral segments filiform, longer than the middle one, spur longer than the ovary, upcurved, anther large, apiculate, the cells parallel, the tubes short, upcurved; glands of pollinia orbicular; rostellum erect, slender, stigmatic process clavate.

Very rare in Horsley hills in Chittoor district.

Fl. and Fr. : July–August.

Status: Low risk.

28. **Habenaria roxburghii** Nicolson in Saldanha and Nicolson, Fl. Hassan Dist. 834. 1876. *Orchis roxburghii* Pers. Syn. Pl. 2: 503. 1807. *Habenaria platyphylla* (Willd.) Spreng. in L. Syst. Veg. (ed. 16.). 690. 1826; FBI 6: 140. 1890; Fischer 3: 1470. 1928.

Family: Orchidaceae

A terrestrial herb, up to 40 cm tall. Leaves 3–6, appressed to the earth, sessile, broadly elliptic or orbicular. Flowers white, in terminal raceme.

In most of the districts. Fl. and Fr.: July–October.

Status: Vulnerable.

29. **Heterostemma deccanense** (Talb.) Swarup & Manglay in J. Linn. Soc. 101: 249–259. 1989. *Oianthus deccanensis* Talb. For. Fl. Bombay Sind. 2: 260. 1911.

Family: Asclepiadaceae

A wiry climber with pubescent branches; latex milky. Petiole 3–8 cm long. Leaves opposite, thick, obovate, to 20 × 17 cm, obtuse or rounded, apex acute to acuminate, glabrous, basal nerves 7 pairs, spreading. Flowers small, in subsessile to peduculate cymes arising between the petioles; calyx lobes ovate, obtuse, 2 mm long, corolla ovoid-urceolate, lobes 2.5 mm long, coronal scales rounded. Follicles 2, divergent, 7–10 cm long, seeds 8–12, imbricated into two rows.

Rare in moist deciduous forests in Maredumilli in East Godavari district and Sukkumamidi in Khammam district (K.N. Reddy *et al.*, 2001). Vern.: Peddapala joola, Pedda Gana joola.

Fl. and Fr.: July–December.

Status: Critically endangered.

30. **Hybanthus vatsavayii** C.S. Reddy in J. Econ. Tax. Bot. 25: 219. 2001.

Family: Violaceae

Erect herb, branched from base, up to 20 cm tall, stem pale yellow puberulous. Leaves linear to linear-lanceolate, 2–4 × 0.1–0.5 cm, base alternate, margin entire to distantly serrate, apex acute-apiculate. Flowers axillary, solitary, inconspicuous. Capsule 3-lobed, puberulous, seeds 5–8 per capsule, ovoid-ellipsoid

Rare in Khammam, Warangal and Nalgonda districts in crevices of rocks.

Fl. and Fr.: August–December.

Status: Not evaluated.

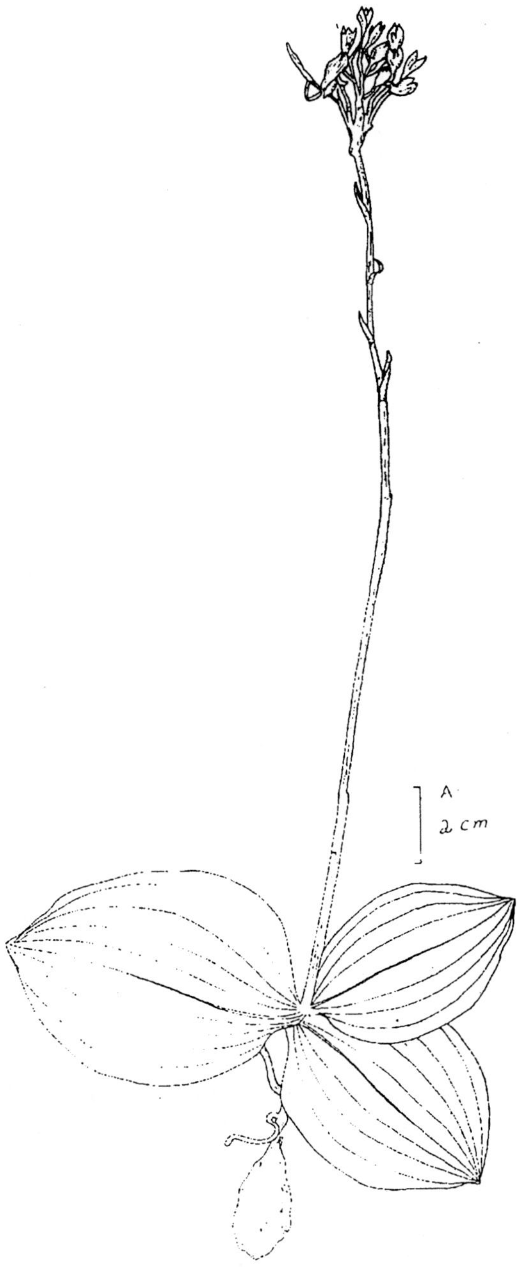

Fig. 10. *Habenaria roxburghii* Nicolson: A. Twig.

31. Iseilema venkateswarluii Satyavathi in J. Bombay Nat. Hist. Soc. 65: 667. 1969.

Family: Poaceae (Gramineae)

An annual prostrate grass. Culms up to 40 cm long, rooting at nodes, round and purple or white in colour. *Involucral spikelets*: 6.5–7.5 mm, oblong-acute, markedly umbonate at base; pedicels 1.25 mm, densely ciliate on the outer margins. Stamens 3. *Sessile spikelet*: Central 5.5 mm of which 3.5 mm are elliptic and the reminder beak; in some racemes beak longer than lower elliptic portion. Glumes ovate-elliptic; lower floret empty; upper floret female; lemma lanceolate; upper lemma bidentate, awned. *Pedicelled spikelet*: present. Glumes 5.5 mm long. Stamens 3, pedicels 4–4.5 mm long.

Rare, in Lam farm area in Guntur district.

Status: Not evaluated.

32. Oianthus disciflorus Hook.f. Fl. Brit. India 4: 49. 1883.

Family: Asclepiadaceae

Twining undershrub, latex milky, young stems pubescent. Leaves ovate, 2–8 × 1.5–3 cm, glabrous, base rounded or subcordate, entire, apex acuminate; petioles to 2.3 cm long. Flowers pale violet with pink tinge inside, in small subsessile lateral cymes arising from between the petioles. Follicles pointed, linear, pale brown when ripe. Seeds comose.

Very rare in bushes in Dhone reserve forest in Yerramalais and Diguvametta in Nallamalais in Kurnool district.

Fl. and Fr.: July–October.

Status: Vulnerable.

33. Parahyparhenia bellarensis (Hack.) Clayton in Kew Bull. 27: 448. 1972. *Andropogon bellarensis* Hack. in Flora 68: 124. 1885; FBI 7: 201. 1896. *Heteropogon bellarensis* (Hack.) Fischer in Gamble Fl. Madras 3: 1744. 1934.

Family: Poaceae

An annual, tufted grass. Culms up to 60 cm tall, slender, terete, glabrous. Leafsheaths terete, glabrous; blades 7.5–20 × 0.3–0.4 cm, involute, filiform, glabrous. Raceme solitary, 5–8 cm long, slender, partially sheathed; joints ciliate. *Sessile spikelets*: 4–7 cm, glabrous; callus bearded. Lower glume linear-oblong, bidentate, 2-keeled, at the apex with a deep median groove on the dorsal surface; upper glume narrowly oblong-lanceolate; lower lemma ovate-oblong, upper lemma reduced to a hyaline base of an awn, 2-lobed. Caryopsis ovoid. *Pedicelled spikelet*:

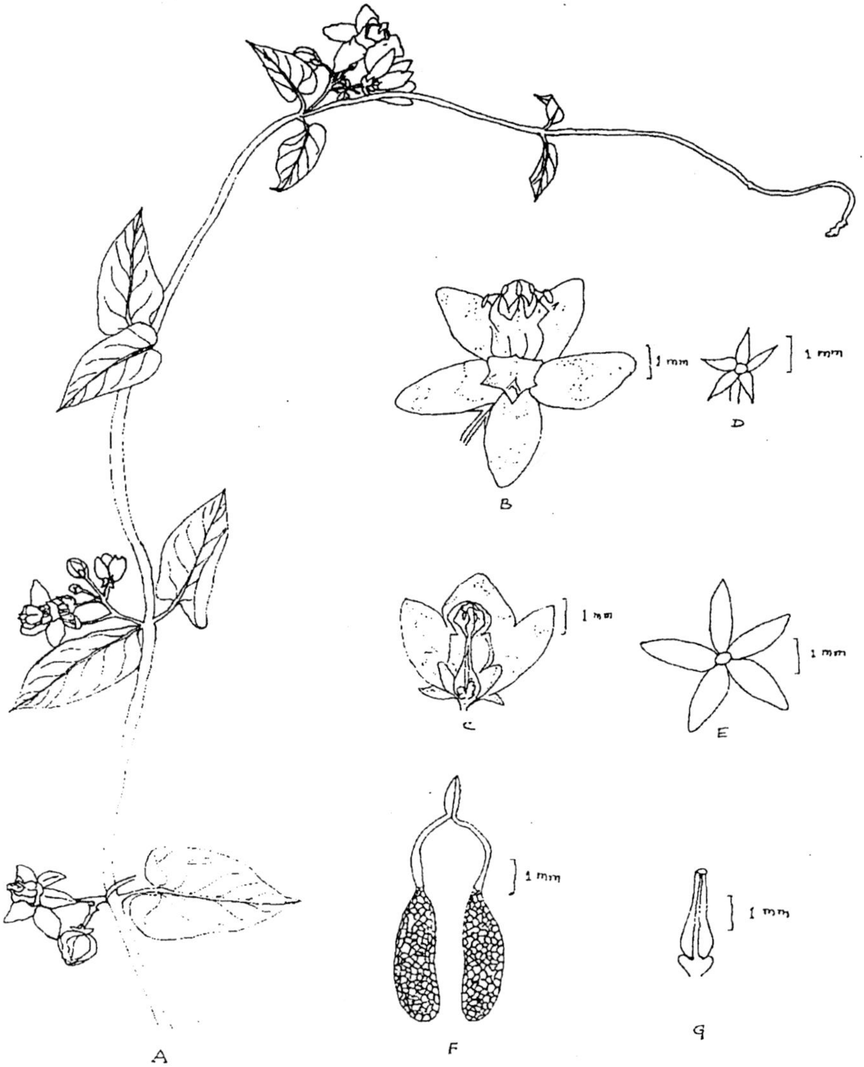

Fig. 11. *Oianthus disciflorus* Hook.f.: A. Twig, B. Flower, C. L.s.of flower, D. Calyx split open, E. Corolla split open, F. Pollinia, G. Pistil.

Oblong, glabrous; callus short; lower glume linear-lanceolate; upper glume linear.

This species was collected by Robert Wight from Gooty hills. Possibly extinct. Hackel (1885) description of *Andropogon bellarensis* was based on Wight 2321 collected from Gooty fort hill, Anantapur district. No specimen at CAL, DD or MH. Isotype is at Kew. Fischer mentions of another collection by Campbell as well from the same locality. Efforts made by us for its collection now proved futile.

34. **Phoenix robusta** (Becc.) Becc. & Hook. f. Fl. Brit. India 6: 427, 1892; Fischer 3: 1559. 1931.

Family: Arecaceae (Palmae)

A plant with trunk 5–7 m tall, clothed and appearing tessellated from the spirally arranged small leaf sheaths. Leaves 1–2 m long, compound, leaflets fascicled, quadrifarious, shining. Spathes narrow, male flowers dense, female flowers distant. Fruit oblong ellipsoid, brown.

Through out the state.

Fl. and Fr.: January–April.

Status: Vulnerable.

35. **Polyzygus tuberosus** Dalz. in Hook. Kew Journ. Bot. 2: 260. 1850; FBI 2: 698. 1879; Gamble 1: 568. 1919.

Family: Apiaceae (Umbelliferae)

Perennial, glabrous herbs, up to 45 cm tall, root tuberous, stem erect. Leaves ternately and pinnately decompound, the segments pinnatifid. Flowers small, white, in terminal and axillary compound umbels, 3–8 rayed. Fruit small, dorsally compressed, ovoid.

In Visakhapatnam district.

Fl. and Fr.: August–December.

Status: Vulnerable.

36. **Pterocarpus santalinus** L.f. Suppl. Pl. 318. 1781; FBI 2: 239. 1876; Gamble 1: 385. 1918. *Ligonium santalinum* (L.f.) O.Kuntze, Rev. Gen. Pl. 1: 193. 1891.

Family: Fabaceae

Large deciduous tree, up to 20 m tall, bark blackish brown, divided into rectangular plates by deep vertical and horizontal cracks, exuding deep red juice when cut. Leaves 3-foliolate, to 15 cm, leaflets ovate-orbicular or oblong, 4.5–10 × 3.2–6 cm, coriaceous, broadly elliptic, shiny, glabrous, base rounded, apex obtuse with slightly emarginate, margin entire, petiole to 4 cm, petiolule to 6 mm. Racemes axillary, 7 cm long, flowers to 2 cm across, pedicel to 5 mm; calyx tomentose, tube 5 mm, lobes ovate; corolla yellow, standard ovate, 2.1 × 1.5 cm, wings 1.8 × 0.8 cm, keels 1.4 × 0.35 cm, staminal sheath 4 × 3 mm, filaments 5 mm; monadelphous (10); ovary stipitate, 7 mm, style 4 mm. Pod obliquely orbicular, 3–4 cm in diameter, gradually narrowed into short stalk or stipe, brown; seeds 1 or 2, reddish brown.

Endemic to hills of Chittoor, Cuddapah, Kurnool, Nellore and adjoining areas.

Fl.: April -June, Fr.: Ripen till February or March. Vern.: Tel.: Rakta chandanam, Yerra chandanam; Tam: Chensandanam, Semmaram. English: Red Sanders.

Status: Endangered.

37. **Rauvolfia serpentina** (L.) Benth. ex Kurz. For. Fl. Brit. Burma 2: 171. 1877; FBI 3: 632. 1882; Gamble 2: 807. 1923. *Ophioxylon serpentinum* L. Sp. 1043. 1753.

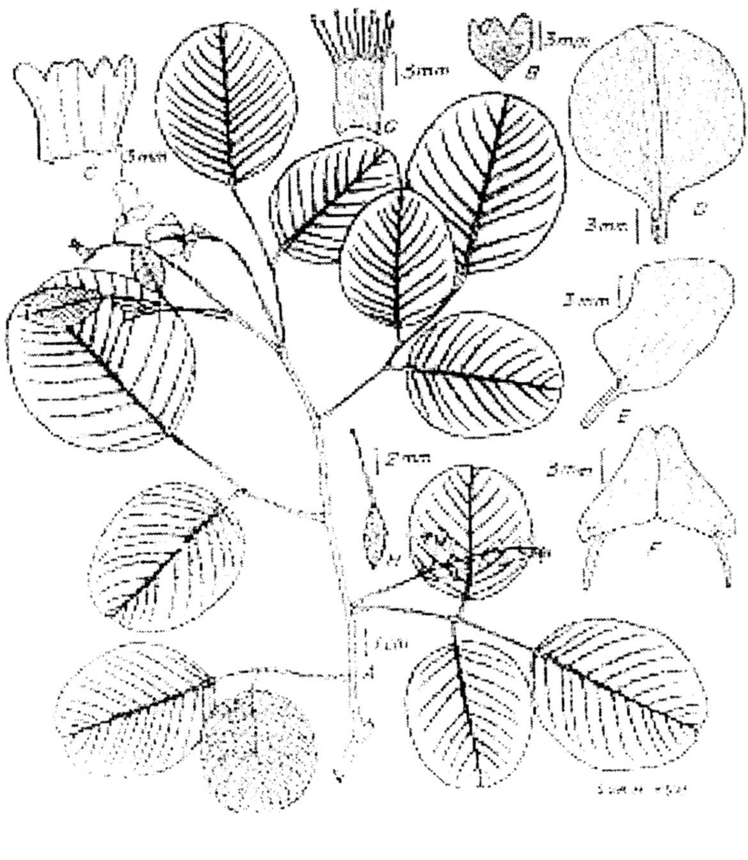

Fig. 12. *Pterocarpus santalinus* L.f.: A. twig, B. Calyx, C. Calyx split open, D. Standard petal, E. Wing petal, F. Keel petals, G. Staminal column, H. Pistil

Family: Apocynaceae

Small erect herb, up to 2 m tall. Leaves in whorls of 3, lanceolate, 10–15 × 3–5 cm, glabrous, bright green above, pale beneath, base attenuate, apex acute to shortly acuminate; main nerves 8–10 pairs; petiole to 2 cm. Flowers white, often tinged with violet, in many-flowered irregular corymbose cymes. Drupes connate except at top, 0.7 cm across, purplish black when ripe.

North coastal Andhra Pradesh.

Fl and Fr.: March–May.

Status: Endangered.

38. **Scyphiphora hydrophyllacea** Gaertn. Fruct. 3: 91. t. 196. 1805; FBI 3: 125. 1880; Gamble 2: 620. 1921.

Family: Rubiaceae

A small glabrous tree; branchlets swollen at the nodes. Leaves coriaceous, ovate-oblong, 2.5–6 × 1.5–3.5 cm, base acute, margin entire, apex obtuse, glabrous, shining; petiole 1 cm; stipules interpetiolar, short, connate into a sheath around the stem. Flowers in axillary fascicles, white. Drupe oblong, 2 × 0.5 cm, 8–10-grooved.

Rare in Sacremento light house area in East Godavari district.

Fl and Fr.: March–September.

Status: Rare

39. **Shorea tumbaggaia** Roxb. Fl. India 2: 617. 1832; FBI 1: 306. 1874; Gamble 1: 83. 1915. *Vatica tumbaggaia* (Roxb.) Wight & Arn. Prodr. Fl. Ind. Orient 84. 1834.

Family: Dipterocarpaceae

Deciduous tree, up to 20 m tall; bark dark brown, thick, rough, longitudinally fissured; wood brown, branchlets tomentose. Leaves thick-coriaceous, 6–18 × 3–10 cm, ovate, glabrous on both surfaces, apex acuminate, margin entire, base cordate or subcordate; secondary nerves ca. 9 pairs, 7 mm apart, petiole 1.5–4.5 cm. Flowers white, fragrant, in 15–20 cm long terminal panicles. Capsule yellowish white, belly (seed part) 2 cm long, ovoid, pubescent; wings unequal, spathulate, to 5 cm long, nerved.

Scarce, chiefly in hilly areas and to some extent in gregarious patches, mostly on the upper slopes and crests of hills and on plateaus of Seshachalam and Veligonda hills up to 1000 m.

Fl.: March–April; Fr.: June–August. Vern.: Tel: Thamba Jalari, Thamba, Guggilam, Nalladammara, Nallaguggilam.

Andhra Pradesh and Tamil Nadu–endemic.

Status: Critically endangered.

40. Vanilla wightiana Lindl. Gen. Sp. Orchid. 436. 1840; FBI 6: 90. 1890; Fischer 3: 1451. 1928.

Family: Orchidaceae

A scandent leafless herb, climbing by means of aerial roots from the nodes; stems fleshy, green. Scape arising from the nodes, 2–3 cm long, 3–5-flowered. Flowers cream coloured with a pale green tinge, 3 cm across, sepals oblong-ovate, obtuse or subacute at apex; petals oblong, obtuse at apex, lip trumpet-shaped; disc barbate-hairy.

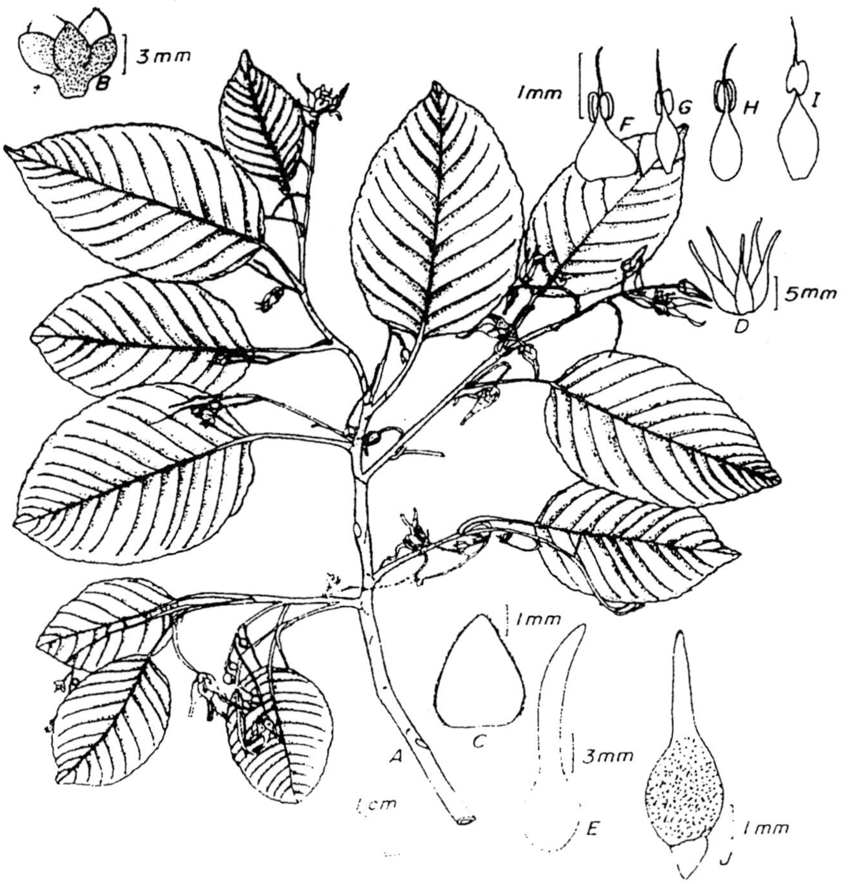

Fig. 13. *Shorea tubaggaia* Roxb.: A. Twig, B. Calyx, C. Sepal, D. Corolla, E. Petal, F-I, Stamens, J. Pistil.

Capsule cylindric, to 15 × 1.2 cm, smooth.

In forests of Rajavommangi around Surampalem in East Godavari district. Also in Chittoor district.

Fl. and Fr.: March–April.

Status: Rare.

41. Cyathea gigantea (Wall. ex Hook.) Holtt. in Gard. Bull. Settl. 8: 318. 1935. *Alsophila gigantea* Wall. ex Hook. Sp. Fil. 1: 53. 1884.

Family: Cyatheaceae

Trunk about 10 cm in diameter, about half a metre high with persistent swollen bases of stipes, bearing crown of fronds at the apex; trunk densely covered by scales; scales mixture of narrow, oblong, about 4 × 1 mm and lanceolate, about 8 × 1 mm scales, uniformly dark brown, glossy, acuminate, entire. Stipes tufted, up to 60 × 2 cm, brown, glossy, abaxially rounded, adaxially grooved, densely scaly at the swollen base, glabrous above; elliptic to linear, about 5 × 1 mm streaks distributed all over the surfaces of stipes and rachis. Lamina bipinnate, deltoid, about 140 × 90 cm, primary pinnae about 10 pairs, spreading, alternate, distinctly stalked, oblong lanceolate, up to 740 × 15 cm, apex acuminate, base truncate; secondary pinnae about 25 pairs, catadromous, spreading, alternate, about 2 cm apart, shortly stalked, oblong-lanceolate, about 12 × 2 cm, apex acuminate, base truncate, margin usually lobed, 4 mm to the costa, crenate, lobes about 10 pairs, oblong or broadly deltoid, about 4 × 5 mm, apex rounded, margin crenate; costa and costules well distinct below, slightly distinct above; veins of the adjacent lobes reaching the side of the sinus independently, lamina with pale brown, soft, acicular hairs densely distributed on the adaxial side of the secondary rachis and costa in addition to few small, narrow, linear, pale brown scales. Sori exindusiate medianly superficial on each ultimate veinlet, nearly at the middle of the lowest veinlet and progressively closer to base in the anterior ones till on the apical pair the sorus is at or very close to base of veinlet, forming an inverted 'V' on each lobe, the extreme apical odd veinlet not bearing sori, sporangia numerous, compact; paraphyses mingled with sporangia.

Terrestrial along shaded stream banks in Visakhapatnam, East and West Godavari districts.

India: Throughout in hilly regions.

World: Nepal, China, Burma, Thailand, Sri Lanka, Bangladesh.

Status: Low risk.

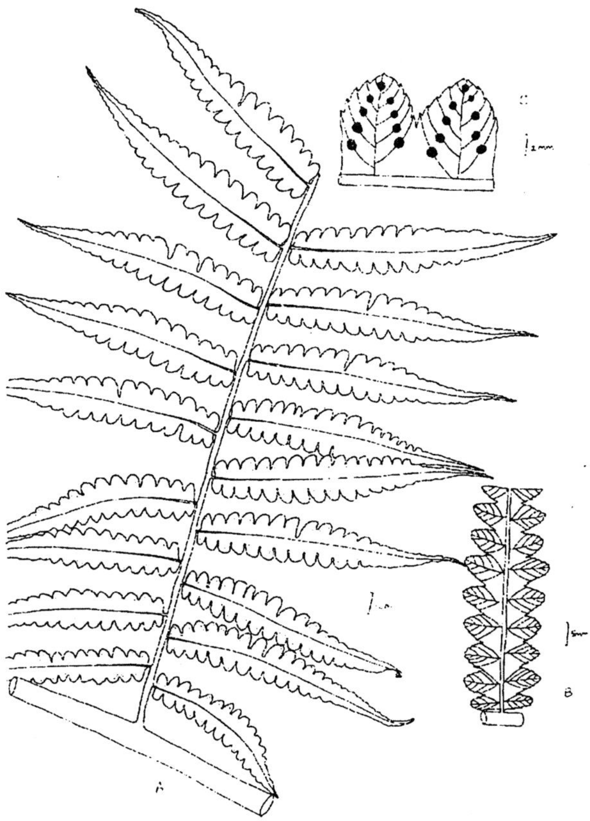

Fig. 14. *Cyathea gigantea* (Wall. ex Hook.) Holttum: A. Primary pinna, B. Basal part of secondary pinna enlarged showing venation and sori, C. Two lobes of secondary pinna enlarged showing venation and sori

42. Cyathea nilgirensis Holttum in Kew Bull. 19: 468. 1965.

Family: Cyatheaceae

Trunk about 3m high, with crown of fronds at the apex; scales densely covering the younger fronds, up to 6 × 1 mm, lanceolate, dark brown, apex hair pointed, margin fimbriate. Stipes about 80 × 0.2 cm, swollen at the base, hairy. Lamina bipinnate, oblong-lanceolate up to 150 × 80 cm, acute; pinnae about 10 pairs, alternate, distinctly petiolate, up to 12cm apart, 40 × 23 cm, oblong, lanceolate, acuminate, base truncate; basal few pairs slightly reduced; pinnules up to 15 × 3 cm, acuminate, base cuneate, margin pinnatifid up to the costa; ultimate lobes up to 15 pairs, alternate, oblong, 1.5 × 0.3 cm, falcate, apex rounded, margin crenate; basal most acroscopic lobe larger than others; veins distinct well below, slightly above, all the veins except few distal pairs forked once, free, reaching the margin; pinnae pale green below, dark green above; texture herbaceous, pale brown, ovate, appressed scales distributed on lower surface of the costa and on lower half of the costules below; ferrugineous hairs also distributed densely on upper surface of the costa and costules. Sori seated at the vein forks of the lower half of the segments spherical, exindusiate, paraphyses intermingled with sporangia; spores trilete.

Terrestrial along shaded stream banks or along shaded roadsides in Sunkarimetta, Galikonda of Visakhapatnam and Tadepalli of East Godavari district.

India: South India, endemic.

Status: Rare.

43. Cyathea spinulosa Wall. ex Hook., Sp. Fil. 1: 25 t. 12c. 1844.

Family: Cyatheaceae

Rhizome trunk like, 20–30 cm thick, covered with a thick mat of clinging roots and persistent leaf bases and the younger regions covered with dark-brown paleae which are 30 × 3 mm, Fronds crowded with stipes, up to 80 cm long, 2 cm thick, dark brown, shining, muricate, shallowly grooved on the adaxial side and spinose having prominent woody spines (larger ones 5mm) on the abaxial side. Lamina ovate-oblong, 2 × 1m, 2-pinnate–3-pinnatifid with the rachis similar to stipe, bearing 12 pairs of primary pinnae of which the larger ones are ovate-oblong, up to 60 × 20 cm; secondary pinnae 20–25 pairs per primary pinnna, narrowly oblong, 10 × 2 cm, subsessile and with the margin pinnatisect into oblong, subfalcate subsessile lobes up to 7 × 3 mm, with sharply toothed margin having a narrow thickened revolute cartilaginous edge: secondary rachis spinulose below, muricate and shallowly grooved above and bearing small paleae as on stipe and rachis; lamina thin and herbaceous. Midrib of secondary pinnae raised on both surfaces, bearing dense acicular hairs on the upper and pale small bullate paleae on the lower; midrib of pinnae lobes similar and bear 8–10 pairs of free veinlets which are forked near the base and all veinlets merging with the cartilaginous margin of lamina. Sori restricted to lateral veinlets in basal half of the lobes, seated at forking of the lateral veinlets, crowded when mature and protected by an urn-shaped thin membraneous indusium which breaks irregularly into a few broad flaps when sporangia mature.

Grows on the banks of streams and rivulets in semi-shaded localities in Visakhapatnam, East and West Godavari districts.

India: Throughout in hilly regions.

World: Nepal, Bhutan, China, Burma, Thailand, Japan.

Status: Low risk.

44. Equisetum ramosissimum Desf., Fl. Atlant. 2: 398. 1800.

Family: Equisetaceae.

Rhizome long creeping, subterranean, branched, dark brown, terete with ridges and furrows, up to 5 mm thick with nodes and internodes, bearing fibrous roots and aerial stem, internodal length up to 5 cm,

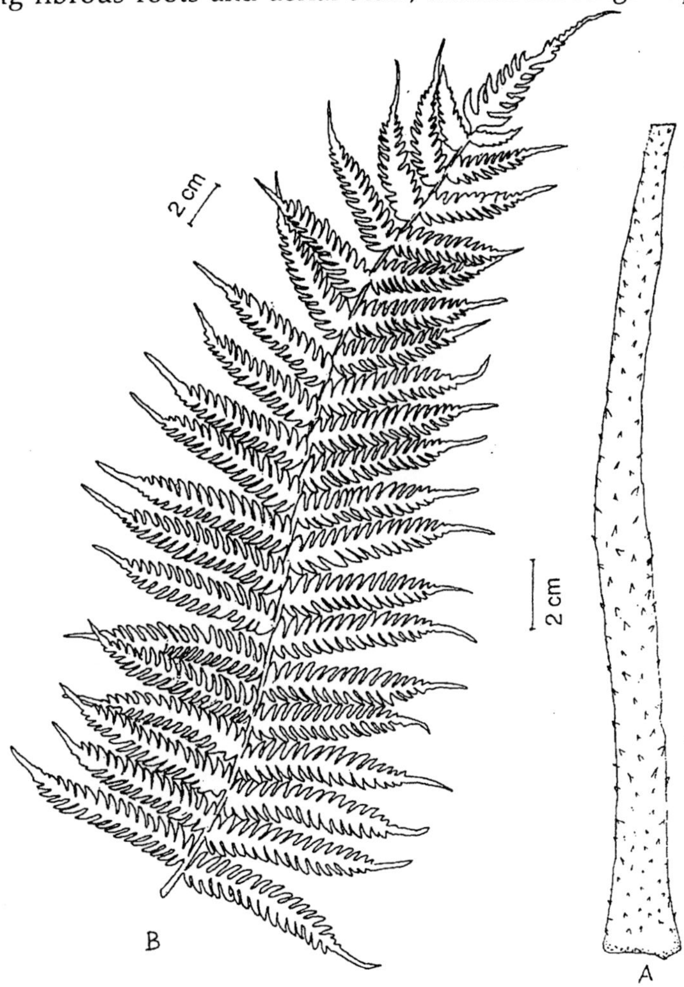

Fig. 15. *Cyathea spinulosa* Wall. ex Hook.: A. Basal part of stipe, B. Part of primary pinna.

each node bears up to 0.7 cm tall cylindrical sheath. Aerial stem up to 2 m tall and distinguished into nodes and internodes; inter nodal region tubular, up to 10cm long, surface with many ridges and furrows, nodal region not distinguishable, concolorous with inter nodal part, the dark brown upper edge of the nodal sheath apparently looking like the node; lateral branches one to five, around the nodes at the base of the nodal sheath, erect or pendent, up to 50 × 0.2 cm, similar to the main stem; leaves scale-like borne on the upper edge of the nodal sheath, up to 6 × 1 mm, narrowed, dark and opaque with a thin, membranaceous border at the basal half, margin incised at the distal half; aerial stem pale green, glabrous, coriaceous.

Very rare species. Terrestrial as clumps on roadside in Araku, Machigedda R.F. and Kotnigedda in Visakhapatnam district. It also occurs at Sunkesula in Kurnool district.

India: North India, Tamil Nadu.

World: Tropical.

Status: Low risk.

45. Gnetum ula Brongn.in Duperr. Voy. Monde 7: 121. 1829; Fischer 3: 1885. 1936. *G. funiculare* Sm. ex Wight Ic. t. 1955. 1853 non Bl. 1834. *G. scandens* auct. non Roxb.; FBI 5: 642. 1881 p.p.; Fischer 3: 1392. 1928.

Family: Gnetaceae

Large dichotomously branched woody climber, branchlets terete, nodes dilated, bark thick, brown, rough. Leaves elliptic, elliptic-ovate, elliptic-oblong or ovate, 7.5–21 × 4.5–10 cm, entire, shortly acuminate, glabrous, midrib stout, base rounded or very obtuse; petiole 0.7–1.8 cm. Flowers monoecious (or dioecious). Male spike 2.5–5 cm long, peduncled, in 3-chotomous panicles. Female inflorescence not branched; fruiting spike up to 25 cm long. 'Fruit' orange when mature, ellipsoid, 2.5–4.5 × 1.5–2.5 cm.

Rare along valleys in Araku.

Fl.: April – May, Fr.: October-November.

Rare in East Godavari and Visakhapatnam districts.

Status: Rare.

46. Psilotum nudum (L.) P. Beauv., Prod. Fam. Aetheog. 106, 112. 1805. *Lycopodium nudum* L., Sp. Pl. 2: 1100. 1753.

Family: Psilotaceae

Rhizome short creeping, branched, up to 2 mm thick, dark brown, densely covered by dark brown, slender unicellular hairs; roots absent. Stem erect or arching, main branched stem up to 13 × 0.2 cm, more or less tetragonal with few grooves and ridges, iso or aniso dichotomously branched, up to five times at the distal part to form semi orbicular of about 10 × 8 cm shoot system; scale leaves regularly arranged in alternate pairs throughout the branches, pale green, up to 1 × 0.5 mm, narrowly triangular to lanceolate; stem dark green, glabrous all over except the presence of scale leaves. Sporangia trilocular, synangia borne at axis of scale leaves, sessile, up to 1mm in diameter, dehisced by vertical slit.

Rare in Galikonda hill slopes, Chintapalli in Visakhapatnam. Talakona and Papavinasanam in Chittoor district.

India: Throughout.

World: Burma, Malaysia.

Status: Rare.

Threatened Medicinal Plants Listed by Sudhakar Reddy *et al.* (2001).

1. *Aegle marmelos* (L.) Corr.
   Family: Rutaceae
   Status: Vulnerable
2. *Aphanamixis polystachya* (Wall.) Parker
   Family: Meliaceae
   Status: Vulnerable
3. *Aristolochia tagala* Cham.
   Family: Aristolochiaceae
   Status: Vulnerable
4. *Arthraxon depressus* Stapf ex Fischer
   Family: Poaceae
5. *Arundinella setosa* Trin. var. *lanifera* Fischer
   Family: Poaceae
   Status: Rare
6. *Balanophora dioica* Royle
   Family: Balanophoraceae
   Status: Rare
7. *Baliospermum montanum* (Willd.) Muell.-Arg.
   Family: Euphorbiaceae
   Status: Vulnerable

8. *Bambusa arundinacea* (Retz.) Roxb. var. *gigantea* Bahadur
   Family: Poaceae
   Status: Rare

9. *Barleria gibsoni* Dalz.
   Family: Acanthaceae
   Status: Vulnerable

10. *Barleria prattensis* Sant.
    Family: Acanthaceae
    Status: Vulnerable

11. *Begonia trichocarpa* Dalz.
    Family: Begoniaceae
    Status: Endangered

12. *Cadaba fruticosa* (L.) Druce
    Family: Capparaceae
    Status: Near threatened

13. *Cayratia pedata* (Lam.) Gagnep.
    Family: Vitaceae
    Status: Rare

14. *Celastrus paniculatus* Willd.
    Family: Celastraceae
    Status: Vulnerable

15. *Ceropegia bulbosa* Roxb.
    Family: Asclepiadaceae
    Status: Endangered

16. *Ceropegia candelabrum* L. var. *biflora* (L.) Ansari
    Family: Asclepiadaceae
    Status: Endangered

17. *Ceropegia candelabrum* L. var. *candelabrum*
    Family: Asclepiadaceae
    Status: Endangered

18. *Ceropegia hirsuta* Wight and Arn.
    Family: Asclepiadaceae
    Status: Endangered

19. *Ceropegia juncea* Roxb.
    Family: Asclepiadaceae
    Startus: Endangered

20. *Ceropegia pusilla* Wight & Arn.
    Family: Asclepiadaceae
    Status: Endangered

21. *Chlorophytum arundinaceum* Baker
    Family: Liliaceae
    Status: Endangered

22. *Chlorophytum glaucum* Dalz.
    Family: Liliaceae
    Status: Vulnerable

23. *Commelina hirsuta* (Wight) Clarke
    Family: Commelinaceae
    Status: Rare

24. *Corymborkis veratrifolia* (Koen.) Smith
    Family: Orchidaceae
    Status: Vulnerable

25. *Crotalaria filipes* Benth.
    Family: Fabaceae
    Status: Endangered

26. *Crotalaria longipes* Wight & Arn.
    Family: Fabaceae
    Status: Endangered

27. *Crotalaria rigida* Heyne ex Roth
    Family: Fabaceae
    Status: Rare

28. *Curcuma amada* Roxb.
    Family: Zingiberaceae
    Status: Vulnerable

29. *Curcuma decipiens* Dalz.
    Family: Zingiberaceae
    Status: Vulnerable

30. *Curcuma inodora* Blatt.
    Family: Zingiberaceae
    Status: Vulnerable

31. *Curcuma pseudomontana* Grah.
    Family: Zingiberaceae
    Status: Vulnerable

32. *Curcuma zedoaria* (Christm.) Roscoe
    Family: Zingiberaceae
    Status: Near Threatened

33. *Cycas circinalis* L.
    Family: Cycadaceae
    Status: Threatened

34. *Dalbergia horrida* (Dennst.) Mabb.
    Family: Fabaceae
    Status: Vulnerable

35. *Decalepis hamiltonii* Wight & Arn.
    Family: Asclepiadaceae
    Status: Endangered

36. *Dendrobium ovatum* (Willd.) Kranz.
    Family: Orchidaceae
    Status: Vulnerable

37. *Dicanthium maccanii* Blatter
    Family: Poaceae
    Status: Vulnerable

38. *Diospyros candolleana* Wight
    Family: Ebenaceae
    Status: Vulnerable

39. *Drosera burmanii* Vahl
    Family: Droseraceae
    Status: Vulnerable

40. *Drosera indica* L.
    Family: Droseraceae
    Status: Vulnerable

41. *Embelia tsjeriam-cottam* (Roem. & Schult.) DC.
    Family: Myrsinaceae
    Status: Endangered

42. *Entada pursaetha* DC.
    Family: Fabaceae
    Status: Vulnerable

43. *Eulalia wightii* (Hook.f.) Bor
    Family: Poaceae
    Status: Rare

44. *Eulophia nuda* Lindl.
Family: Orchidaceae
Status: Endangered

45. *Fagonia cretica* L.
Family: Zygophyllaceae
Status: Vulnerable

46. *Fumaria indica* (Hassk.) Pugsley
Family: Fumariaceae
Status: Endangered

47. *Gardenia gummifera* L.f.
Family: Rubiaceae
Status: Vulnerable

48. *Gardenia resinifera* L.f.
Family: Rubiaceae
Status: Near threatened

49. *Gloriosa superba* L.
Family: Liliaceae
Status: Vulnerable

50. *Gymnema sylvestre* (Retz.) Schult.
Family: Asclepiadaceae
Status: Near threatened

51. *Habenaria longicorniculata* Grah.
Family: Orchidaceae
Status: Vulnerable

52. *Hedychium coronarium* Koen.
Family: Zingiberaceae
Status: Near threatened

53. *Heracleum ringens* Wall. ex DC.
Family: Apiaceae
Status: Vulnerable

54. *Heteropogon contortus* (L.) Beauv. ex Roem. & Schult. var. *polystachyos* (Roxb.) Deshpande
Family: Poaceae
Status: Indeterminate

55. *Holostemma ada-kodien* Schult.
Family: Asclepiadaceae
Status: Vulnerable

56. *Iphigenia pallida* Baker
Family: Liliaceae
Status: Vulnerable

57. *Kingiodendron pinnatum*
Family: Caesalpiniaceae
Status: Vulnerable

58. *Knema attenuata* (Wall. ex Hook.f. & Thoms.) Warb.
Family: Myristichaceae
Status: Near threatened

59. *Moringa concanensis* Nimmo ex Gibs.
Family: Moringaceae
Status: Vulnerable

60. *Mucuna monosperma* Wt.
Family: Fabaceae
Status: Vulnerable

61. *Nervilia aragoana* Gaud.
Family: Orchidaceae
Status: Near threatened

62. *Operculina turpethum* (L.) S.Manso (=*Merremia turpethum* (L.) Shah & Bhat)
Family: Convolvulaceae
Status: Endangered

63. *Oropetium roxburghianum* (Steud.) Philips
Family: Poaceae
Status: Rare

64. *Oroxylum indicum* (L.) Vent.
Family: Bignoniaceae
Status: Endangered

65. *Panicum fischeri* Bor
Family: Poaceae
Status: Rare

66. *Persea macrantha* (Nees) Kosterm.
Family: Lauraceae
Status: Endangered

67. *Piper longum* L.
Family: Piperaceae
Status: Endangered

68. *Piper nigrum* L.
Family: Piperaceae
Status: Near threatened

69. *Pseudarthria viscida* (L.) Wight & Arn.
Family: Fabaceae
Status: Vulnerable

70. *Pterocarpus marsupium* Roxb.
Family: Fabaceae
Status: Vulnerable

71. *Pterospermum xylocarpum* (Gaertn.) Santapau & Wagh
Family: Sterculiaceae
Status: Near threatened

72. *Pueraria tuberosa* DC
Family: Fabaceae
Status: Vulnerable

73. *Rhaphidophora pertusa* (Roxb.) Schott.
Family: Araceae
Status: Vulnerable

74. *Rubia cordifolia* L
Family: Rubiaceae
Status: Vulnerable

75. *Salacia oblonga* Wall. ex Wight & Arn.
Family: Celastraceae
Status: Endangered

76. *Santalum album* L.
Family: Santalaceae
Status: Vulnerable

77. *Saraca asoca* (Roxb.) de Wilde
Family: Caesalpinaceae
Status: Vulnerable

78. *Schrebera swietenioides* Roxb.
Family: Oleaceae
Status: Vulnerable

79. *Smilax zeylanica* L.
Family: Liliaceae
Status: Vulnerable

80. *Terminalia arjuna* (Roxb. ex DC.) Wight & Arn.
    Family: Combretaceae
    Status: Near threatened

81. *Tinospora sinensis* (Lour.) Merril
    Family: Menispermaceae
    Status: Vulnerable

82. *Trichosanthes anaimalaiensis* Bedd.
    Family: Cucurbitaceae
    Status: Critically endangered

83. *Trichosanthes cucumeriana* L.
    Family: Cucurbitaceae
    Status: Data deficient

84. *Tripogon jacquemontii* Stapf
    Family: Poaceae
    Status: Rare

85. *Tripogon wightii* Hook.f.
    Family: Poaceae
    Status: Rare

86. *Vitex trifolia* L.
    Family: Verbenaceae
    Status: Near threatened

87. *Wendlandia angustifolia* Hook.f.
    Family: Rubiaceae
    Status: Endangered.

REFERENCES

Nayar, M.P. and A.R.K. Sastry. 1987. Red Data book of Indian Plants. Bot. Surv. India.

Sudhakar Reddy, Ch., K.N. Reddy and S.N. Jadhav. 2001. Threatened (Medicinal) Plants of Andhra Pradesh. Medicinal Plants conservation centre, Hyderabad.

Suryanarayana, B. and A. Sreenivasa Rao. 2002. Flora of Nellore district, Andhra Pradesh. Gurudev Prakashan, Shrirampur.

CHAPTER 8

FLORA OF VISAKHAPATNAM STEEL PLANT TOWNSHIP AREA

*M.N.V. Peddi Naidu and B. Hanumantha Rao**

Introduction

Located on the beautiful coastline of the Bay of Bengal, Visakhapatnam city hosts a number of major industries. Out of the several major industries the Steel Plant has not only changed the destiny of several thousands of people here, but it has also thrown open new vistas of development. Visakha Steel township comprises of eleven sectors with 8000 quarters surrounded by good vegetation. Of the total area of 8827 hectares covered by the Steel Plant nearly 3600 hectares of land is under the shade of good vegetation. Visakhapatnam city approximately lies between the East longitudes 83° 15′–83° 22′ and North latitude 17° 14′–17° 46′. The city has on the eastern side the shore of Bay of Bengal. The popular Visakhapatnam Steel Plant is situated at around 32 km distance from the city between Visakhapatnam and Anakapalli on the National Highway No. 5. The township has the Kanithi colony (Gajuwaka-west) and the shore of Bay of Bengal on the eastern side. On the northern side lie Kanithi Balancing Reservoir, Visakha Steel General Hospital, Aganampudi approach road, Aganampudi and Kurmannapalem. On the southern side are present the Gangavaram beach and the Visakha Steel Plant.

The lands are mostly composed of broken ground, rocky soil dotted with big black boulders and a curious red earth. The Visakhapatnam Steel Plant township is pleasant like most of the sea side places in the

* Department of Botany, Andhra University, Visakhapatnam – 530003

tropics. The nearby hills on the eastern side of Gangavaram sea shore deflects the debilitating long shore winds and turns it into sea breeze. North-East monsoon (December–February), Hot weather season (March-May), South-West monsoon (June to September) and retreating South-West monsoon season (October to December) are the four conventional seasons of the year. The rainfall is strongly seasonal and is generally confined to the South-West monsoon. About 70% of the yearly rainfall is received during this season and the average varies from 500 to 1000 mm. In the remaining seasons, rainfall, if any, is mainly due to thunder storms and cyclones. The average temperatures are high and uniform and the mean annual temperature is approximately 30°C and the highest mean temperature is reached in the month of May. The relative humidity varies from 60–90% with some diurnal fluctuations in the various seasons mentioned above. Cyclones are not infrequent. Their activity is mainly initiated in the Bay of Bengal and these are associated with hurricane winds often producing widespread rains.

The location under study belongs to the coastward low level zone and composed almost of Khondelite with bands of other rock types such as laterites, pegmatites, gravites, leptyrites and charnokite series. The vegetation is basically of the dry deciduous type with some evergreen species here and there.

The vegetation of the Coromandel coast was previously described by Roxburgh (1795–1819) in his monumental work entitled 'Plants of the coast of Coromandel'. Later Gamble and Fischer (1915–1936) also made references to the plants of Visakhapatnam in their 'Flora of Madras Presidency'. A precise and up to date data on the Floa of Visakhapatnam city were given by Venkatswarlu *et al.* in the year 1972 in their excellent volume "The Flora of Visakhapatnam'. Subsequently Subba Rao (1977, 1993) also added some more information on this flora covering the interior forest areas of Visakhapatnam district. In the recent years the overall development of the city along with the extension of its limits has the impact on wild vegetation also. Especially consequent to the establishment of Visakhapatnam Steel Plant along with its beautiful Township having well designed sectors and people of varied cultures, an obvious change could also be seen in the vegetation of this particular area. Since no critical enumeration data is available so far for this area, an attempt has been made here especially to provide an authentic and detailed up to date data on the Flora of Visakhapatnam Steel Plant Township area.

During the course of this study on the Flora of Visakhapatnam Steel Plant Township area a total collection of 1102 field numbers, comprising around 217 species and 177 genera or Angiosperms were collected.

The families are arranged by following Bentham and Hooker's

(1862–1883) system of classification. Species in a family are arranged in alphabetical sequence. It is followed by the habit, flower colour and field number (PN: Peddi Naidu).

Systematic Enumeration

MAGNOLIACEAE

Michelia champaka Linn.
Evergreen tree. Flowers yellow or orange. PN 216, 271, 310, 381.

ANNONACEAE

Annona reticulata Linn.
Small tree. PN 372, 422, 503

Annona squamosa Linn.
Small tree. Flowers pale yellow. PN 24, 56, 116, 172, 574, 606.

Artabotrys odoratissimus R.Br.
Climbing shrub. Flowers greenish yellow. PN 252, 316, 442, 466.

Cananga odorata (Lam.) Hook.f. & Thoms.
Medium sized to large tree. Flowers greenish. PN 201, 286, 772, 844.

Polyalthia longifolia Benth. & Hook.f. ex Hook.f. & Thoms.
Evergreen tree. Flowers yellowish green. PN 74, 145, 197, 467.

PAPAVERACEAE

Argemone mexicana Linn.
Prickly annual herb. Flowers yellow. PN 443, 473, 514, 993.

CAPPARACEAE

Capparis zeylanica Linn.
Large climbing thorny shrub. Flowers white or rose. PN 122, 177, 683, 732.

Cleome gynandra Linn.
Erect annual herb. Flowers white. PN 347, 363, 468, 532, 862.

BIXACEAE

Bixa orellana Linn.
Small evergreen tree. Flowers white or purplish. PN 21, 81, 125, 571.

POLYGALACEAE

Polygala chinensis Linn.
Annual herb. Flowers yellow. PN 183, 221, 264, 713.

CARYOPHYLLACEAE

Polycarpaea corymbosa (Linn.) Lamk.
Annual erect herb. Flowers white. PN 323, 383, 432, 900, 914.

PORTULACEAEAE

Portulaca quadrifida Linn
Diffuse herb. Flowers yellow. PN 84, 131, 444, 469, 534.
Talinum cuneifolium Willd.
Erect, succulent herb. Flowers purple. PN 204, 262, 313, 771, 821.

CLUSIACEAE (GUTTIFERAE)

Calophyllum inophyllum Linn.
Medium sized tree. Flowers fragrant. PN 223, 793.

MALVACEAE

Abelmoschus esculentus (Linn.)Moench.
Annual herb. Flowers yellow. Cultivated. PN 374, 423, 472, 522, 943.
Abutilon crispum (Linn.) Medik.
Diffuse annual herb. Flowers yellow. PN 225, 283, 753, 842.
Abutilon indicum Sweet
Shrub. Flowers yellow. PN 166, 445, 470, 535.
Ceiba pentandra (Linn.) Gaertn.
Tall tree. Flowers white. PN 34, 134, 185, 582.
Gossypium herbaceum Linn.
Shrub. Flowers yellow. Cultivated. PN 385, 942.
Hibiscus cannabinus Linn.
Erect herb. Flowers yellow. PN 393, 431, 531, 913.
Hibiscus micranthus Linn.f.
Erect herb. PN 162, 234, 384, 471, 536.
Hibiscus rosa-sinensis Linn.
Shrub. Flowers crimson. Ornamental PN 1, 41, 96, 217, 382, 412, 460.
Pavonia zeylanica Cav.
Erect undershrub. Flowers pink. PN 111, 181, 281, 361, 474.
Sida acuta Burm.
Undershrub. Flowers yellow. PN 25, 46, 174, 233.
Sida cordifolia Linn.
Undershrub or herb. Flowers yellow. PN 163, 212, 261, 475.
Sida veronicaefolia Lamk.
Prostrate or trailing herb. Flowers pale yellow. PN 133, 161, 241, 446.

Thespesia populnea (Linn.) Soland. ex Corr.
Medium sized tree. Flowers bright yellow. PN 272, 321, 373, 835, 891.

STERCULIACEAE

Melochia corchorifolia Linn.
Spreading herb. Flowers pink. PN 89, 273, 392, 413, 624.

Sterculia foetida Linn.
Medium sized tree. Flowers dark brown or dull orange. PN 123, 171, 682, 742.

Waltheria indica Linn.
Perennial herb. Flowers yellow. PN 302, 353, 477, 871, 941.

TILIACEAE

Corchorus aestuans Linn.
Annual erect herb. Flowers yellow. PN 391, 414, 452, 501, 944.

Corchorus fascicularis Lam.
Annual erect herb. Flowers yellow. PN 222, 272, 312, 761, 834.

Corchorus olitorius Linn.
Erect undershrub. Flowers yellow. PN 151, 231, 265, 722.

Corchorus trilocularis Linn.
Annual erect herb. Flowers yellow. PN 7, 102, 153, 551, 652, 733.

Triumfetta rhomboidea Jacq.
Erect undershrub. Flowers yellow. PN 101, 211, 301 411, 478, 684.

ZYGOPHYLLACEAE

Tribulus terrestris Linn.
Annual prostrate, villous herb. Flowers yellow. PN 121, 173, 479, 539, 671.

OXALIDACEAE

Oxalis corniculata Linn.
Diffuse herb. Flowers yellow. PN 71, 371, 433, 612, 916.

RUTACEAE

Citrus aurantifolia (Christm.) Swingle
Tree. Fruit globose, yellow or green when ripe. Planted. Tel.: Nimma. PN 73, 126, 176, 622, 725.

Citrus limon (Linn.) Burm.f.
Tree. Planted.

Citrus medica Linn. var. **limon** Linn.
Tree. Fruit globose or ovoid. Planted. Tel: Dabba. PN 326, 892.

Feronia limonia (Linn.) Swingle
Deciduous armed tree. Flowers pale green. Tel.: Velaga. PN 61, 603.
Glycosmis mauritiana (Lamk.)Tanaka
Erect shrub. Flowers white. Tel: Gonji. PN 182, 226, 734.
Murraya koenigii (Linn.) Spreng.
Small tree. Leaves aromatic. Tel: Karivepa, Karepaku. PN 314, 364, 883.

SIMAROUBACEAE

Ailanthus excelsa Roxb.
Deciduous tree. Flowers polygamous. Tel: Peddamanu. PN 461, 1002.

MELIACEAE

Azadirachta indica A.Juss.
Deciduous tree. Flowers white. Tel: Vepa. PN 3, 51, 152, 552, 701, 921.
Melia azedarach Linn.
Small tree. Flowers pinkish blue. Tel.: Turakavepa. PN 4, 202, 553, 751.
Sweitenia mahagoni Jacq.
Evergreen tree. Flowers white-pale green. PN 103, 651.

RHAMNACEAE

Zizyphus mauritiana Lamk.
Small tree. Flowers greenish. Tel.: Regi, Gangareni. PN 105, 154, 203, 251, 801.
Zizyphus oenoplia Mill.
A thorny large climber. Flowers white or yellowish. Tel: Pariki. PN 253, 802.

VITACEAE

Cissus quadrangularis Linn.
Fleshy tendril climber. Tel: Nalleru. PN 106, 155, 205, 662, 702.

SAPINDACEAE

Cardiospermum halicacabum Linn.
Tendrilar climbing herb. Flowers white. Tel: Buddakakara. PN 206, 303, 401, 480, 754.
Dodonaea viscosa (Linn.)Jacq.
Evergreen shrub. Flowers greenish yellow. PN 107, 676.
Sapindus emarginatus Vahl.
Moderate sized tree. Tel.: Kunkudu. PN 254, 304, 803, 852.

ANACARDIACEAE

Anacardium occidentale Linn.
Tree. Tel: Jeedimamidi. PN 26, 47, 355, 576, 911, 1030.

Buchanania angustifolia Roxb.
Medium sized tree. Tel: Sarapappu. PN 207, 755.

Mangifera indica Linn.
Large evergreen tree. Flowers white. Tel: Mamidi. PN 27, 48, 354, 922, 955.

MORINGACEAE

Moringa oleifera Lamk.
Medium sized tree. Flowers white. Tel: Munaga, Mulaga. PN 356, 404, 923.

FABACEAE

Abrus precatorius Linn.
Wiry woody climber. Flowers pinkish. Tel: Guruvinda. PN 108, 305, 357, 656.

Alysicarpus monilifer DC.
Prostrate annual herb. Flowers pink. Tel.: Amera. PN 358, 405, 917, 957.

Arachis hypogaea Linn.
Annual herb. Cultivated. Tel.: Verusanaga. PN 109, 156, 657, 703.

Clitoria ternatea Linn.
Climber. Flowers blue or white. Tel.: Nallaghentana. PN 28, 306, 359, 482.

Crotalaria laburnifolia Linn.
Erect shrub. Flowers yellow. Tel.: Peddagiligicha. PN 29, 52, 579, 602.

Crotalaria verrucosa Linn.
Annual hereb. Flowers bluish or white. PN 30, 53, 110, 447, 1033.

Dalbergia latifolia Roxb.
Deciduous tree. Flowers white. Tel.: Viruguducheva. PN 208, 756.

Dalbergia sisso Roxb.
Small tree. Flowers creamy. PN 5, 554.

Desmodium triflorum (Linn.)DC.
Trailing prostrate herb. Flowers pink or purplish. Tel.: Muntamandu. PN 54, 112.

Indigofera hirsuta Linn.
Undershrub. Flowers pink or rose. PN 307, 448, 540, 854.

Pongamia pinnata (Linn.) Pierre
Deciduous tree. Flowers pinkish white. Tel.: Kanuga. PN 55, 607.

Pterocarpus santalinus Linn.f.
Tall deciduous tree. Planted. Tel.: Raktachandanamu, Yerrachandanamu. PN 57, 113, 608.

Stylosanthes mucronata Willd.
Annual stiff undershrub. PN 6, 115, 158, 665, 706.

Tephrosia maxima Pers.
Perennial herb. Flowers purple or pale red. PN 31, 209, 255, 406, 581, 1034.

Tephrosia procumbens Ham.
Procumbent perennial herb. Flowers white. PN 8, 42, 117, 187, 592.

Tephrosia purpurea (Linn.) Pers.
Annual or perennial under shrub. Flowers purple rose or red. Tel: Vempali. PN 2, 58, 118, 214, 462, 609, 1070.

Tephrosia villosa (Linn.) Pers.
Erect, silky villous undershrub. Flowers bluish white or pale red. Tel: Nuguvempali. PN 256, 309, 400, 805, 856.

Zornia diphylla (Linn.) Pers.
Annual prostrate or spreading herb. Flowers white or cream. PN 257, 311, 375, 524, 910.

CAESALPINIACEAE

Bauhinia purpurea Linn.
Evergreen tree. Flowers purple or rose. PN 9, 557.

Bauhinia vahlii Wt. & Arn.
Large woody climber. Flowers cream coloured. Tel.: Adda. PN 315, 858.

Caesalpinia bonduc (Linn.) Roxb.
Armed scandent shrub. Flowers yellow. Tel.: Gachakaya. PN 119, 668.

Cassia alata Linn.
Large shrub or small tree. PN 317, 362, 859, 927.

Cassia auriculata Linn.
Undershrub. Flowers yellow. Tel.: Tangedu. PN 215, 258, 318, 360, 485, 1090.

Cassia fistula Linn.
Medium sized tree. Flowers bright yellow. Tel.: Rela. PN 259, 319, 808, 920.

Cassia marginata Roxb.
Small tree. Flowers red or rose coloured. Tel.: Sima rela. PN 59, 120, 159, 610.

Cassia occidentalis Linn.
Annual herb or undershrub. Flowers yellow or orange yellow. Tel.: Kasivinda. PN 10, 43, 196, 260, 453, 593, 1073.

Cassia siamea Lamk.
Small tree. Flowers yellow. Tel.: Simatangedu. PN 367, 408, 937, 963.

Cassia tora Linn.
Annual herb. Flowers yellow. Tel.: Tantepu. PN 124, 164, 213, 707, 759.

Delonix elata Gamble
Medium sized tree. Flowers yellow. Tel.: Vatanarayana, Sunkesula. PN 263, 332.

Delonix regia (Boj.) Raf.
Large tree. Flowers red. Tel.: Turayi chettu. PN 11, 60, 97, 297, 613.

Parkinsonia aculeata Linn.
Perennial shrub. Flowers yellow. Tel.: Simatumma. PN 266, 811.

Peltophorum pterocarpum (DC.) Baker ex Heyne
Large evergreen tree. Flowers yellow. Tel.: Kondachinta. PN 375, 930.

Poinciana pulcherrima Linn.
Prickly shrub. Flowers red or orange. Tel.: Ratnagandhi. PN 320, 863.

Tamarindus indica Linn.
Large tree. Flowers yellow. Tel.: Chinta. PN 62, 128, 614, 708.

MIMOSACEAE

Acacia nilotica (Linn.) Willd. ex Del. ssp. **indica** (Benth.) Brenan
Small tree. Flowers yellow. Tel.: Nalla tumma. PN 12, 560, 616.

Albizia lebbeck Linn.
Deciduous tree. Flowers white. Tel: Dirisana. PN 129, 679.

Mimosa pudica Linn.
Spreading herb. Flowers pink. Tel.: Peddanidrakanti. PN 369, 505, 964, 1052.

Pithecellobium dulce (Roxb.) Benth.
Armed tree. Tel.: Simachinta. PN 130, 681.

Samanea saman (Jacq.) Merril
Large tree. Flowers yellowish. PN 368, 939.

COMBRETACEAE

Quisqualis indica Linn.
Climbing shrub. Flowers red. Ornamental. Tel.: Rangoon malle. PN 220, 766.

Terminalia catappa Linn.
Large tree. Tel.: Badamu. PN 167, 715.

MYRTACEAE

Eucalyptus camaldulensis Dehnh.
Large tree. Flowers white. Planted. Tel: Jamoil. PN 64, 135, 617, 685.

Psidium guajava Linn.
Medium sized tree. Flowers white. Tel.: Jama. PN 399, 407, 940.

Syzygium cumini (Linn.) Skeels
Evergreen tree. Flowers white. Tel.: Neredu. PN 132, 680.

LYTHRACEAE

Lawsonia inermis Linn.
Shrub. Flowers white. Tel.: Gorinta. PN 370, 946.

Punica granatum Linn.
Large shrub or small tree. Flowers red. Cultivated. Tel.: Danimma. PN 378, 947.

PASSIFLORACEAE

Passiflora foetida Linn.
Tendril climber. Flowers white. Tel.: Tellajumiki. PN 65, 618.

CUCURBITACEAE

Benincasa hispida (Thunb.) Cogn.
Climbing herb. Fruit large. Cultivated. Tel.: Budida gummadi. PN 451, 935, 1001.

Coccinia cordifolia (Linn.) Cogn.
Tendrillar climbing herb. Flowers white. Cultivated and wild. Tel.: Kakidonda (wild), Donda (cultivated). PN 136, 686.

Cucumis sativus Linn.
Annual prostrate creeper. Flowers yellow. Cultivated. Tel.: Dosakaya. PN 219, 767.

Cucurbita maxima Duch.
Prostrate herb. Cultivated. Tel.: Tiyyagummadi, Gummadi. PN 322, 865.

Lagenaria siceraria (Molina) Standl.
Climber. Cultivated.Tel.: Sorakayi. PN 66, 619.

Luffa acutangula (Linn.) Roxb.
Extensive climber. Flowers unisexual. Cultivated. Tel.: Bira. PN 267, 838.

Momordica charantia Linn.
Climbing herb. Flowers yellow. Tel.: Kakara. PN 13, 561.

Trichosanthes anguina Linn.
Climber. Cuitivated. Tel.: Potla. PN 268, 814.

CACTACEAE

Opuntia dillenii Haw.
Leafless spinous shrub. Flowers yellow. Tel.: Nagajemudu. PN 324, 377, 506.

AIZOACEAE

Mollugo disticha Ser.
Annual erect herb. PN 14, 562.

Mollugo nudicaulis Lamk.
Annual herb. Leaves all radical. PN 409, 454, 960, 1054.

Mollugo pentaphylla Linn.
Erect annual herb. Flowers white. Tel.: Verri chatarasi. PN 410, 965.

Trianthema portulacastrum Linn.
Prostrate herb. Flowers white. Tel.: Galijeru. PN 227, 760.

APIACEAE (UMBELLIFERAE)

Centella asiatica (Linn.) Urban.
Trailing herb. Flowers reddinsh or pink. Tel.: Saraswati aku. PN 325, 867.

RUBIACEAE

Borreria articularis (Linn.f.) F.N.Williams
Prostrate herb. Flowers pale pink or rose. Tel.: Madana. PN 32, 67, 583, 620.

Hamelia patens Jacq.
Evergreen shrub. Flowers reddish yellow or reddish orange. PN 269, 815.

Ixora coccinea Linn.
Large shrub. Flowers bright scarlet red. Ornamental. Tel.: Ramabanamu. PN 33, 137, 584, 687.

Morinda angustifolia Roxb.
Small tree. Flowers white. Tel.: Togaru. PN 229, 768.

ASTERACEAE (COMPOSITAE)

Acanthospermum hispidum DC.
Annual erect herb. Flowers yellow. PN 270, 327, 868, 933.

Ageratum conyzoides Linn.
Annual erect herb. Flowers pale blue or white. PN 72, 168, 542, 709, 1091.

Eclipta prostrata (Linn.) Linn.
Prostrate or erect herb. Flowers white or pale blue. PN 23, 224, 440, 573, 744.

Parthenium hysterophorus Linn.
Annual erect herb. Flowers white. PN 100, 169, 441, 510, 563, 649, 1055.

Sphaeranthus indicus Linn.
Annual aromatic herb. Flowers pink or purple. PN 329, 508, 870, 1056.

Tridax procumbens Linn.
Hispid prostrate herb. Flowers yellow. Tel.: Gaddi chamanti. PN 68, 543, 585, 770, 1037.

Vernonia cinerea (Linn.) Less.
Annual, erect herb. Flowers pink, rose or lilac. Tel.: Gharitikamini. PN 36, 49, 544, 717, 1038.

Xanthium strumarium Linn.
Herb. Heads monoecious. PN 75, 625.

PLUMBAGINACEAE

Plumbago zeylanica Linn.
Perennial shrub. Flowers white. Tel.: Chitramulamu. PN 230, 776.

SAPOTACEAE

Achras zapota Linn.
Evergreen tree. Flowers white. Tel.: Sapota. PN 416, 966.

OLEACEAE

Jasminum angustifolium Willd.
Wiry climbing shrub. Flowers white, often purplish outside. Tel.: Adavimalli. PN 232, 777.

Jasminum multiflorum (Burm.f.) Andr.
Tel.: Virajaji. PN 277, 845.

Jasminum sambac (Linn.) Ait.
Flowers white. Tel.: Malli. PN 380, 948.

APOCYNACEAE

Alstonia scholaris (Linn.) R.Br.
Evergreen tree. PN 418, 967.

Allamanda cathartica Linn.
Shrub. Flowers bright yellow. PN 278, 822.

Catharanthus roseus (Linn.) Don
Undershrub. Flowers pink, red, rose or white. Tel.: Billaganneru. PN 331, 455, 509, 1005.

Nerium indicum Mill.
Large shrub. Flowers rose, red or white. Tel.: Ganneru. PN 279, 511, 817, 1058.

Thevetia peruviana (Pers.) Merr.
Small tree. Flowers golden yellow. Tel.: Suvarna ganneru, Paccha ganneru. PN 69, 626.

Wrightia tinctoria R.Br.
Deciduous tree. Flowers white. Tel.: Ankudu, Tellapala. PN 330, 512, 873, 1060.

ASCLEPIADACEAE

Calotropis gigantea (Linn.) R.Br.
Large shrub. Tel.: Jilledu. PN 333, 456, 513, 875, 1006, 1061.

BORAGINACEAE

Coldenia procumbens Linn.
Prostrate, scabrid herb. Flowers pale yellow. PN 76, 627.

CONVOLVULACEAE

Cuscuta reflexa Roxb.
Twining complete stem parasite. Flowers white. PN 37, 77, 587, 628.

Evolvulus alsinoides Linn.
Prostrate herb. Flowers blue. Tel.: Vishnukranti. PN 38, 78, 138, 588, 688, 1039.

Ipomoea carnea Jacq.
Shrub. Flowers pale rose. PN 170, 235, 490, 547, 718, 778.

Ipomoea kentrocaulos C.B.Clarke
Climber. Flowers white. Tel.: Cheralapilliteega. PN 39, 50, 457, 515, 589, 600.

Ipomoea pentaphylla Jacq.
Climbing or twining herb. Flowers white. PN 458, 1008.

Ipomoea pes-capre Sweet
Creeping herb. Flowers pink. PN 178, 236, 336, 779, 819.

Ipomoea pes-tigridis Linn.
Prostrate herb,. Flowers white. PN 299, 459, 516, 1009.

Ipomoea reptans Poir.
Aquatic herb. Flowers pale purple. Tel.: Tutikuna. PN 1101, 1102.

Merremia tridentata (Linn.) Hallier f.
Prostrate or twining herb. Flowers white. PN 40, 73, 139, 450, 491, 590.

SOLANACEAE

Capsicum annuum Linn.
Annual erect herb. Flowers white. Cultivated. Tel.: Mirapakaya. PN 284, 386, 824, 949.

Cestrum diurnum Linn.
Shrub. Flowers white. Ornamental. PN 420, 970.

Cestrum nocturnum Linn.
Scandent shrub. Flowers white. Ornamental. Tel.: Night queen. PN 419, 460, 969.

Datura metel Linn.
Shrub. Flowers white. Tel.: Ummetta. PN 86, 282, 492, 517, 622, 823.

Lycopersicon esculentum Mill.
Annual weak stemmed herb. Flowers yellow. Cultivated. Tel.: Tomato. PN 300, 387, 850.

Physalis minima Linn.
Annual erect herb. Flowers yellow. Tel.: Kupanti. PN 16, 70, 564, 630.

Solanum melongena Linn.
Cultivated. Tel.: Vanga, Vankaya. PN 179, 495, 720.

Solanum nigrum Linn.
Annual, erect herb. Flowers white. Tel.: Kamanchi. PN 237, 494, 550, 1044.

Solanum surattense Burm.f.
Prostrate perennial herb. Flowers blue or purple. Tel.: Challa mulaga. PN 91, 140, 493, 641, 1043.

BIGNONIACEAE

Kigelia pinnata DC.
Tree. Flowers liver coloured. Planted. PN 79, 633.

Spathodea campanulata Beauv.
Medium sized tree. Flowers bright orange-crimson or vermillon. Planted. Tel.: Patadi. PN 388, 901.

Tabebuia pentaphylla (Linn.) Hemsl.
Tree. Flowers white. Planted. PN 80, 141, 238, 634.

Tecoma stans D.Don
Large shrub. Flowers golden yellow. Planted. Tel.: Swarnaganneru. PN 389, 902.

ACANTHACEAE

Andrographis paniculata (Burm.f.) Wall. ex Nees
Erect herb. Flowers pink or purple. Tel.: Nelavemu. PN 180, 727.

Asystasia gangetica (Linn.) T.Anders.
Erect herb. Flowers white. PN 390, 403, 421, 496, 972.

Justicia procumbens Linn.
Diffuse herb. Flowers purple or pale violet. PN 285, 334, 424, 497, 876, 974.

Justicia prostrata Gamble
Slender prostrate herb. Flowers pale pink. PN 17, 565.

Justicia simplex D.Don
Slender herb. Flowers pale pink or lilac. PN 85, 142, 186, 498, 692.

Ruellia tuberosa Linn.
Erect herb. Flowers deep blue. PN 143, 693.

Thunbergia grandiflora (Roxb. ex Rottl.) Roxb.
Large climber. Flowers blue. PN 18, 566.

VERBENACEAE

Duranta repens Linn.
Shrub. Flower blue or violet. Ornamental. PN 144, 425, 694, 975.

Lantana camara Linn. var. **aculeata** (Linn.) Moldenke
Straggling armed shrub. Flowers pink, orange or white. Tel.: Pulikampa. PN 146, 239, 499, 695, 1049.

Tectona grandis Linn.f.
Large tree. Tel.: Teku. PN 287, 335, 827, 878.

LAMIACEAE

Leucas aspera (Willd.) Spreng.
Annual erect herb. Flowers white. Tel.: Tellatummichettu. PN 426, 976.

Leucas biflora (Vahl) R.Br.
Procumbent annual herb. PN 518, 1065.

Ocimum americanum Linn.
Annual, aromatic herb. Flowers white or pale violet. Tel.: Kukkatulasi. PN 19, 82, 567, 636.

Ocimum sanctum Linn.
Erect undershrub. Flowers pink. Tel.: Tulasi. PN 184, 240, 288, 729, 788.

NYCTAGINACEAE

Boerhavia diffusa Linn.
Spreading herb. Flowers pink. Tel.: Punarnava. PN 337, 500, 885, 1050.

Bougainvillea spectabilis Willd.
Scandent shrub. Ornamental. PN 338, 519, 879, 1066.

Mirabilis jalapa Linn.
Erect herb. Flowers white, red or yellow. Ornamental. Tel.: Chandrakanta. PN 242, 789.

AMARANTHACEAE

Achyranthes aspera Linn.
Erect herb. Flowers greenish white. Tel.: Uttareni. PN 339, 394, 880, 904.

Amaranthus spinosus Linn.
Erect spinous herb. Flowers green. PN 87, 637.

Amaranthus viridis Linn.
Erect herb. Flowers pale green. Tel.: Thotakura. PN 147, 696.

BASELLACEAE

Basella rubra Linn.
Twiner. Flowers pink. Tel.: Bachalikura. PN 188, 736.

POLYGONACEAE

Antigonon leptopus Hook. & Arn.
Climbing shrub. Flowers red, rose or occasionally white. Tel.: Batani teega. PN 148, 697.

EUPHORBIACEAE

Acalypha indica Linn.
Annual erect herb. Flowers unisexual. Tel.: Muripindi. PN 289, 340, 828, 886.

Croton bonplandianum Baill.
Annual erect herb. Flower unisexual. PN 20, 83, 189, 568, 638, 737.

Euphorbia hirta Linn.
Annual erect or spreading herb. Tel.: Reddivarinanubalu. PN 341, 395, 427, 887.

Euphorbia pulcherrima Willd. ex Klotz.
Erect shrub. Bracts bright red. PN 342, 888.

Jatropha curcas Linn.
Deciduous shrub or small tree. Flowers unisexual, yellowish green. Tel.: Pedanepalamu. PN 149, 199, 343, 428, 699.

Manihot esculenta Crantz.
Tuberous tall herb. Tel.: Karrapendalamu. Cultivated. PN 90, 150, 191, 640.

Micrococca mercurialis Benth.
Annual herb. PN 243, 344, 790, 890.

Phyllanthus emblica Linn.
Moderate sized tree. Flowers unisexual. Tel.: Peddausiri. PN 396, 907.

Ricinus communis Linn.
Cultivated. Tel: Amudamu. PN 520, 1067.

Tragia involucrata Linn.
Spring like herb with stinging hairs all over. Tel.: Dulagondi. PN 88, 190, 244.

MORACEAE

Artocarpus heterophyllus Lamk.
Large evergreen tree. Fruit very large, tubercled. Tel.: Panasa. PN 249, 795.

Ficus benghalensis Linn.
Large tree. Tel.: Marri. PN 348, 893.

PANDANACEAE

Pandanus fascicularis Lamk.
A large shrub, stem supported by stilt roots. Flowers in cylindrical spikes enclosed by long yellow, fragrant spathes. Tel.: Mogali. PN 194, 746.

CYPERACEAE

Cyperus kyllinga Endl.
Tufted sedge. PN 436, 988.

POACEAE (GRAMINEAE)

Bambusa arundinacea (Retz.) Roxb.
Tall bamboo. Tel.: Veduru. PN 398, 909.

Cynodon dactylon (Linn.) Pers.
Perennial creeping grass with erect stems. Tel: Garikigaddi. PN 437, 523, 989.

Imperata cylindrica (Linn.) P. Beauv.
Tel.: Darbha. PN 95, 645.

Leptochloa chininsis (Linn.) Nees
Annual grass. Tel: Chipurugaddi. PN 295, 438, 829, 990.

Saccharum spontaneum Linn.
Erect grass. Cultivated. Tel.: Rellugaddi. PN 195, 747.

REFERENCES

Gamble, J.S. and C.E.C. Fischer. 1915–1936. Flora of Presidency of Madras. (3 vols. Rep. ed. 1975). Bot. Surv. India, Howrah.

Roxburgh, W. 1795–1819. Plants of the coast of Coromandel. (3 vols.) London.

Subba Rao, G.V. 1977. Flora of Visakhapatnam district, Andhra Pradesh. Bull. Bot. Surv. India 19: 122–126.

Subba rao, G.V. 1992. Flora of Visakhapatnam District. Ph.D. thesis, Andhra University, Visakhapatnam.

Venkateswarlu, J., P.V.B. Murthy and P.N. Rao. 1972. The Flora of Visakhapatnam. Andhra Pradesh Academy of Sciences, Hyderabad.

INDEX TO SCIENTIFIC NAMES AND SUBJECT INDEX*

* Synonyms are given in italics